REVIEW FOR THE CLEP* GENERAL MATHEMATICS EXAMINATION

Complete review of skills

By
Michael O'Donnell
Robert Floyd

This book is correlated to the video tapes produced by
COMEX Systems, Inc., Review for The
CLEP* General Mathematics Examination
by Robert Floyd ©1998
they may be obtained from

comex systems, inc.
5 Cold Hill Rd., Suite 24
Mendham, NJ 07945

* "CLEP and College Level Examination Program are registered trademarks of The College Entrance Examination Board. These Materials have been prepared by Comex Systems, Inc., which bears sole responsibility for their contents."

Published by

comex systems, inc.

5 Cold Hill Rd.
Suite 24
Mendham, NJ 07945

ISBN 1-56030-176-7

Table of Contents

CLEP* (College Level Examination Program)

CLEP provides a way to determine the level of knowledge you now have in relation to college level material. CLEP does not determine your ability to learn a subject. People tend to have a low evaluation of their ability. There is no way you can determine your present level unless you take the examination. You can save time and money taking these examinations to earn credit.

WHY DID WE WRITE THIS BOOK?

Our firm has conducted many classroom reviews for CLEP General Examinations. Our instructors have assisted thousands of candidates. From this experience we have determined that:

1. In each area there is specific material beneficial for candidates to know.
2. There is a need for a simple-to-follow review book which helps students improve their ability to achieve.
3. It is important for students to become accustomed to the specific directions found on the examination before taking the examination.
4. It is beneficial to develop a systematic approach to taking an objective examination.

This book will help you perform at your highest potential so that you will receive your best score.

The flyers "CLEP COLLEGES" (Listing where you may take the CLEP tests and the colleges that accept CLEP for credit) and "CLEP INFORMATION FOR CANDIDATES" are available free by calling (800) 257-9558, by writing to: CLEP, PO Box 6600, Princeton, NJ 08541-6600 or online at www.collegeboard.com.

CLEP INFORMATION

WHAT IS CLEP GENERAL?

CLEP is a nation-wide program of testing which began in 1965. Today thousands of colleges recognize CLEP as a way students may earn college credit. Each year hundreds of thousands of students take CLEP examinations. The testing program is based on the theory that "**what** a person knows is more important than **how** he has learned it". All examinations are designed and scored by the College Entrance Examination Board (CEEB). The purpose of each examination is to determine whether your current knowledge in a subject will qualify you for credit in that area at a particular college.

There are five general examinations. The subject areas are:

1. English Composition
2. Mathematics
3. Social Science / History
4. Natural Science
5. Humanities

Credits earned through achieving on these examinations replace basic liberal arts credits which are required by many colleges. Each of these general examinations is very broad in coverage. Questions are from the wide range of subjects included in each of the major disciplines. The General CLEP Humanities Examination will include questions related to literature, music and art. Because of the broad coverage in each examination, you are not expected to be knowledgeable in all areas. There will be some questions on all the tests you will not be able to answer.

HOW LONG ARE THE EXAMINATIONS?

Each CLEP General Examination is 1½ hours in length. Each examination is divided into separate timed portions.

HOW MUCH DO THE EXAMINATIONS COST?

Currently, the fee to take each examination is $50.00. They may be taken one at a time or in any combination. (NOTE: Fees change periodically.)

WHERE WILL THE EXAMINATIONS BE TAKEN?

The CEEB (College Entrance Examination Board) has designated certain schools in each state to serve as test centers for CLEP examinations. The same examinations are given at each test center. If you are a member of the armed forces, consult with the Education Services Officer at your base. Special testings are set up for military personnel. A list of test centers may be found at www.collegeboard.com.

WHEN ARE THE TESTS GIVEN?

Most CLEP examinations are administered during the third week of every month except December and February. The test center chooses the day of the week. A few test centers administer the tests by appointment only. Check with the center where you will take the test for specific information. If you are serving with the United States Military, check with the Education Services Officer at your base to find out about the DANTES testing program. You will be given information about testing as applicable to military personnel.

HOW DO YOU REGISTER FOR AN EXAMINATION?

A standard registration form may be obtained from the test center where you plan to take the examination. Many centers require that you register (send registration form and fee for examinations to be taken) a month prior to your selected date.

WHEN WILL SCORES BE RECEIVED?

Most tests taken on the computer will be scored immediately (the English Composition with Essay must be graded first). If you are in the military and are taking the test on paper it will take up to 6 weeks to receive your scores. You may also request that a copy be sent to a college. The score you receive will be a scaled score. CEEB keeps a record of your scores on file for 20 years. You may obtain an additional copy or have a copy sent to a college if you contact:

College Board
PO BOX 6600
ATTN: Transcript Service
Princeton, NJ 08541
800-257-9558

IS IT NECESSARY TO BE ENROLLED IN A COLLEGE BEFORE YOU TAKE AN EXAMINATION?

Each college has established policies regarding CLEP. Check with the school you wish to attend. Many schools do not require enrollment before taking CLEP examinations.

HOW MANY CREDITS MAY BE EARNED?

Each college determines the number of credits that may be earned by CLEP examinations. Most colleges award six credits for achievement on a CLEP General Examination.

HOW ARE THE EXAMS SCORED?

See page VII for a detailed explanation of scoring.

HOW ARE THE SCORES EVALUATED?

The examinations are administered to college students who are taking a course the examination credits will replace. These students do not take the examination for credit. They take it to establish a standard by which your score will be evaluated. Percentile levels of achievement are determined. For example, if you score at the 25th percentile, this would indicate that you achieved as well as the **bottom** 25 percent of those students who took that examination to set the standard.

There is no correlation between the number of questions you answer correctly and the percentile level you achieve. The number will vary from test to test.

CAN THE SAME SCORES EARN A DIFFERENT NUMBER OF CREDITS AT DIFFERENT SCHOOLS?

Yes, different schools may require different levels of achievement. Your scores may earn more credits at one institution than at another. For example: if you achieve at the 25th percentile level, you could earn credit at a school which required the 25th percentile level; you could not earn credit at a school which required a higher level of achievement.

CAN CLEP CREDITS BE TRANSFERRED?

Yes, provided the school to which you transfer recognizes CLEP as a way to earn credit. Your scores will be evaluated according to the new school's policy.

CAN AN EXAMINATION BE RETAKEN?

Many schools allow you to retake an examination if you did not achieve the first time. Some do not. Check your particular school's policy before you retake an examination. Be realistic, if you almost achieved the level at which you could earn credit, do retake the examination. If your score was quite low, take the course it was designed to replace.

IF YOU DECIDE TO RETAKE AN EXAMINATION, six months must elapse before you do so. Scores on tests repeated earlier than six months will be canceled.

HOW MAY I FIND OUT WHAT SCHOOLS ACCEPT CLEP?

There are many schools that recognize CLEP as a way to earn credit. For a free booklet, <u>CLEP Test Centers and Other Participating Institutions</u>, which lists most of them, send your request, name, and address to:

> The College Board
> Box 1822
> Princeton, NJ 08541
>> 800-257-9558
>> or check the list at www.collegeboard.com

HOW TO USE THIS BOOK:

Recommended procedure:

1. Complete the review material. Take the short tests included at the end of the lessons.
2. If you do well on the tests, continue. If you do not, review the explanatory information.
3. After completing the review material, take the practice examination at the back of the book. When you take this sample test, try to simulate the test situation as nearly as possible:
   a. Find a quiet spot where you will not be disturbed.
   b. Time yourself accurately.
   c. Practice using the coding system
4. Correct the tests. Determine weaknesses. Go back and review those areas in which you had difficulty.

HOW THE EXAMINATIONS ARE SCORED

There is no penalty for wrong answers. Your score is computed based on the number of correct answers. When you are finished with the test make sure that every question is answered; however, you don't have to answer the question the first time you see it. If you use the coding system you will greatly increase your score.

THE CODING SYSTEM

Over the years COMEX has perfected a systematic approach to taking a multiple choice examination. This is called the coding system. It is designed to:

1.   Get you through the examination as quickly as possible.
2.   Have you quickly answer those questions that are easy for you.
3.   Prevent time wasted on those questions that are too difficult.
4.   Take advantage of all your knowledge of a particular subject. Most people think they can get credit only by knowing an answer is correct. You can also prove your knowledge by knowing an answer is incorrect. The coding system will show you how to accomplish this.
5.   Get all the help possible by using the recall factor. Because you are going to read the total examination, it is possible that something in question 50 will trigger a thought that will help you answer question 3 the second time you read it.
6.   Have your questions prioritized for the second reading.

HOW THE CODING SYSTEM WORKS

We are now going to make you a better test-taker, by showing all of your knowledge and using your time to the greatest advantage. Managing your time on the exam can be as important as knowing the correct answers. If you spend too much time working on difficult questions which you have no knowledge about, you might not get to some easy questions later that you would have gotten correct. This could cause a significant decrease in you score.

Let us attack some sample questions:

1.   George Washington was:

   a.   the father of King George Washington
   b.   the father of Farah Washington
   c.   the father of the Washington Laundry
   d.   the father of Washington State
   e.   the father of our country

As you read the questions you will eliminate all **wrong** answers:

a. father of King George Washington   NO!
b. father of Farah Washington   NO!
c. father of the Washington Laundry   NO!
d. father of Washington State   NO!
e. the father of our country   YES. LEAVE IT ALONE.

The question now looks like this:

1. George Washington was:

a. ~~the father of King George Washington~~
b. ~~the father of Farah Washington~~
c. ~~the father of the Washington Laundry~~
d. ~~the father of Washington State~~
e. the father of our country

Click on the button next to the correct answer and click Next.

```
00:35
          ○ Answer
          ○ Answer
          ○ Answer
          ○ Answer
          ○ Answer

        Time Review Mark        Next
```

These are the buttons you must know how to use!

You are now finished with this question. Later when you get to the review process this question will be sorted as answered. This will be your signal to not spend any more time with this question. Any time spent will be wasted.

2. Abraham Lincoln was responsible for:

a. freeing the 495 freeway
b. freeing the slaves
c. freeing the Lincoln Memorial
d. freeing the south for industrialization
e. freeing the Potomac River

Go through the answers.

a. freeing the 495 freeway No!
b. freeing the slaves Maybe. Always read full question.
c. freeing the Lincoln Memorial   No!
d. freeing the south for industrialization   Maybe.
e. freeing the Potomac River   No!

The question now looks like this:

2. Abraham Lincoln was responsible for:

~~a.   freeing the 495 freeway~~
b.   freeing the slaves
~~c.   freeing the Lincoln Memorial~~
d.   freeing the south for industrialization
~~e.   freeing the Potomac River~~

Should you guess? You have very good odds of getting this question correct. Pick the choice that you feel is the best answer. Often your first guess will be the best. Before clicking the Next button, click on the Mark box. This will tell you later that you were able to eliminate 3 answers before guessing. Now click on Next to go on to the next question.

3. Franklin Roosevelt's greatest accomplishment was:

a.   building the Panama Canal
b.   solving the Great Depression
c.   putting America to work
d.   organizing the CCC Corps
e.   instituting the income tax

Go through the answers:

a.	building the Panama Canal	No! That was a different Roosevelt.
b.	solving the Great Depression	Maybe. Go on to the next answer.
c.	putting America to work	Maybe. On to the next answer.
d.	organizing the CCC Corps	Maybe. On to the next answer.
e.	instituting the income tax	Maybe. Leave it alone!

The question now looks like this:

3. Franklin Roosevelt's greatest accomplishment was:

    a. ~~building the Panama Canal~~
    b. solving the Great Depression
    c. putting America to work
    d. organizing the CCC Corps
    e. instituting the income tax

Should you answer this question now? Not yet. There might be a question later that contains information that would help you eliminate more of the answers. When you can only eliminate one answer, or none at all, your best course of action is to simply click on ⌐Next⌐. This will bring up the next question.

Now look at another question:

4. Casper P. Phudd III was noted for:

    a. rowing a boat
    b. sailing a boat
    c. building a boat
    d. designing a boat
    e. navigating a boat

Even if you have no idea of who Casper P. Phudd III is, read the answers:

    a. rowing a boat   I do not know.
    b. sailing a boat   I do not know.
    c. building a boat   I do not know.
    d. designing a boat   I do not know.
    e. navigating a boat I do not know.

Since you cannot eliminate any of the answers, simply go on to the next question.

Try another question:

5. Clarence Q. Jerkwater III

    a. sailed the Atlantic Ocean
    b. drained the Atlantic Ocean
    c. flew over the Atlantic Ocean
    d. colored the Atlantic Ocean orange
    e. swam in the Atlantic Ocean

Even though you know nothing of Clarence Q. Jerkwater III, you read the answers.

    a.   sailed the Atlantic Ocean   Possible.
    b.   drained the Atlantic Ocean   No way!
    c.   flew over the Atlantic Ocean   Maybe.
    d.   colored the Atlantic Ocean orange   No way!
    e.   swam in the Atlantic Ocean   Maybe.

The question now looks like this:

5.   Clarence Q. Jerkwater III

    a.   sailed the Atlantic Ocean
    b.   ~~drained the Atlantic Ocean~~
    c.   flew over the Atlantic Ocean
    d.   ~~colored the Atlantic Ocean orange~~
    e.   swam in the Atlantic Ocean

Do you take a guess? Not on the first reading of the answers. Let us wait to see if the recall factor will help. Do not click on an answer, but do click on Mark. Then click on Next to get the next question.

Continue in this manner until you finish all the questions in the section. By working in this manner you have organized the questions to maximize your efficiency. When you finish with the last question click on Review. This brings up the listing of all the questions. They will be listed in numerical order. This is not the way you want to view them. You sorted the questions as you went through them. You want to view the questions sorted. Click on Status. Now the questions are sorted for you. Let's review what each type means:

Answered without a check mark.
You knew the correct answer.

Answered with a check mark.
You eliminated three answers.

Not answered with a check mark.
You eliminated two answers.

Not answered without a check mark.
You could not eliminate more than one answer.

The Second Time Through

Now you are ready to start your way through the test the second time. Where do you have the best chance of increasing your score? This question should always be at the top of your mind. "How do I show the maximum amount of information I know?" The best place to start is with the questions that you had some idea about, but not enough to answer. These are the questions where you could eliminate two answers. They are marked with a check mark. Clicking on Review, and then Status will sort the questions for you. All of the questions that are marked but have not been answered are grouped together for you.

Click on the first one in the group. Reread the question and the answers. Did anything in any of the other questions give you information to allow you to eliminate any answers? If the answer is yes that is great! The coding system has worked. If you eliminated one more answer make your guess between the remaining two. Leave the Mark box checked and click on Review to go back to the question list to choose your next question. What if you now know the correct answer? Mark it, and **remove** the check from the Mark box. This question will now be listed as answered. You will not spend any more time on this question. Click on Review to go back to your list of questions.

What should you do if you were unable to eliminate any more answers. Now you still need to guess. While your odds are not as good as if you had eliminated three answers, you will have a better chance than if you had eliminated no answers. Any time you eliminate answers before guessing means you are making an educated guess. Every educated guess you make has a higher chance of being correct than a random guess. More educated guesses means a higher score. Leave the Mark box checked. This indicates that you were not sure of your answer.

Continue with this process until you finish all the questions in the group with a check mark that were not answered. Which questions should you work on next? It is now time to work on the questions you had the least knowledge about. These are the questions without a check mark that are not answered. Use the same process that you used for the previous set of questions. Can you now figure out the correct answer? If so mark it and check the box. If not eliminate as many answers as you can and then choose your best answer. If you guess make sure you check the mark box. Every time you reread a question there is a chance that it will trigger something in your memory that will help you with this question, or with one of the others.

Be very careful to keep track of time. If it is not diplayed at the top of the screen, make sure you click on the box so that it will be displayed. Do not think of the clock as your enemy. It is your friend. It keeps you on your task and keeps you moving efficiently through the test.

When you only have 5 minutes left, make sure that you have every question answered. Remember a blank space counts the same as a wrong answer. If you go through and make an educated guess at all the questions, you will get a better score than if the questions were left blank. Even if you randomly guess you should end up with one correct answer out of every five. Every correct answer will increase your score. While you are guessing, make sure you check the mark box, so that you know you guessed on that question. This allows you to review that question later if time permits.

You are now finally at the point where you only have two types of questions left, those where you knew the correct answer and those where you guessed at the answer. All of the questions are now answered. Does this mean it is time to stop? Not if you want to get your highest score. All of the questions on which you took educated guess have a check mark. Keep working on those problems. Do not waste time looking at any questions that do not have a check. You knew the correct answer and are done with them.

By using the coding system you will move quickly through the test and make sure that you see every question. It also allows you to concentrate your efforts on your strongest areas.

Practice the system while you do your exercises and tests. You can use a similar system with a piece of scratch paper. Put an "A" next to questions as you answer them. Put a check mark next to a question to refer back to it. Then use the system to go back through the test. The system is easy to master and will be an invaluable tool in your test-taking arsenal.

You have now completed the portion of this book which is designed to improve your test-taking ability. When you do the practice exercises and take the sample test, use these techniques you have just learned.

You can use the coding system on any multiple choice exam. This will not only increase your score on that test, but it will also make you more comfortable with using the system. It has been demonstrated many times that the more comfortable you are when you are taking a test the higher your score will be.

SOME BASICS FOR THE TEST DAY

1. Get to the examination location early. If you are taking the examination at a new location - check out how to get there **before** the day of the examination.

2. Choose a seat carefully.
   a. In a large room, choose a quiet corner. If possible, sit facing a wall.
   b. If you go with a friend, do not sit together.

3. Stay with your usual routine. If you normally skip breakfast, do so on the test day as well.

4. If you do not understand the proctor's directions, ask questions.

5. Do not quit. Keep going over questions you were not able to answer the first time. You may work anywhere in each section. Beat the examination, do not let it beat you!

6. If you cannot answer a question, code it and go on to the next. Do not spend a lot of time on one question unless you have already finished the rest of that section. Go through each section and do the easiest questions first, then go back to the difficult ones.

7. **Be sure** you understand the directions for **each** type of test **BEFORE you take the examination**. Not understanding the directions can cause you to lose valuable time when you are taking the actual test.

8. Remember to use the coding system.

9. If you are unfamiliar with how to use a mouse, try to get some practice. Most libraries have computers where you can practice. If you have to learn how to use the mouse at the test site you are putting yourself at a severe disadvantage.

CLEP Review—Mathematics

The purpose of this manual is to help you review for the mathematics you will find on the CLEP examination. By careful study of the explanations and the sample problems you will review mathematics that you have already learned, continue the study of those topics, and be introduced to new areas of mathematics that are part of the CLEP examination. The sample test questions are very similar to the question types you will see on the actual examination. Through practice on the sample test you should be able to recognize those problems which are easier for you and those which are more difficult. This will result in the highest degree of efficiency for your time during the examination. In addition, many test-taking hints will be given to make the examination go easier.

It is very important for you to understand that you are not expected to answer every question on the examination. Some may be too difficult for you. The problems in the examination are not in order of difficulty. Therefore, it is especially important that you make an attempt to get through every problem. There may be "easy" problems at the end of the examination. If you spend unnecessary time on problems you cannot solve, you will lose the opportunity to gain valuable points on problems you never reach. Also, there is often a recall factor in these tests. Something you read in a later problem may give you the clue to an earlier problem.

If you prepare well, you will be confident of performing your best. Begin as soon as possible to prepare by finding a regular time to study with no distractions. A half-hour or 45 minutes of concentrated study will accomplish more than an hour with breaks and distractions. Do not let math anxiety be your downfall. Very little in the examination is beyond your understanding.

OUTLINE OF CLEP EXAMINATION

The examination consists of approximately 60 questions to be answered in 90 minutes. This gives you an average of one and a half-minutes to answer each question. However, some of the questions are more difficult and require much more time, while others require much less time. Therefore, it is important to recognize the types of problems that you know require more time and those that can be answered quickly. This will keep you from wasting time on problems that you have less chance of answering correctly and thus give yourself more time for problems you can solve.

The examination consists of the following topics and the approximate number of problems:

modern mathematics (sets)	6 questions
logic	6 questions
number system	12 questions
functions and graphs	12 questions
probability and statistics	15 questions
other algebra topics	9 questions
TOTAL	60 questions

These are topics usually covered in college mathematics courses for non-mathematics majors. In modern mathematics and logic you will be using some new symbols to replace words which express mathematical concepts. Problems about the number system test your understanding of the different kinds of numbers in our system. An important concept, which leads to more difficult areas, is functions. The explanations in this section will introduce the basic concepts and familiarize you with the function problems you will encounter. Probability and statistics will utilize many familiar words in the mathematical meaning. Other topics include logarithms, complex numbers, and other number bases. These are some of the more difficult topics, and you should not spend too much time on these until you have a grasp of the other areas.

Some of the multiple-choice questions have a "double" set of answers. Statements are given after the problem followed by 4 answers to choose from. Choose the answer that includes the combination of statements that make the question correct.

EXAMPLE: Which of the following is (are) always odd?

        I.   The product of 3 consecutive integers (except zero)

        II.   The product of 3 odd numbers and 1 even number

        III.   The sum of 3 consecutive odd integers

A. I only

B. II only

C. III only

D. I, II, and III

In this type of problem, cross out each answer which contains an incorrect statement. In this example, examining the first possibility, we see that statement I must be even; therefore, you can cross out choices A and D since both contain statement I. Statement II is also even; therefore, you can eliminate choice B. That leaves only choice C. This type of question often lends itself to the elimination of several choices. If one of the first statements can be eliminated, then so can all answers with that statement's number.

Mathematical Symbols

=	equals
≠	does not equal
>	greater than
≥	greater than or equal to (can also be written: $\geqq$)
<	less than
≤	less than or equal to (can also be written: $\leqq$)
( )	parentheses, used to group expressions
∧	and
∨	or
||	is parallel to
⊥	is perpendicular to
⊂	is a subset of
•	multiplication dot, as in x • y
÷	division symbol
∠	angle
→	implies
π	pi; the ratio between the circumference and diameter of a circle; approximately equal to $\dfrac{22}{7}$ or 3.14
%	percent
∈	belongs to, or is a member of a set
:	ratio symbol
~	in logic exercises, this symbol means "not" (i.e. if not Q then not P, written as ~Q → ~P)
√	square root (radical) symbol

Properties of Numbers

COMMUTATIVE

It doesn't matter in what order we add or multiply two numbers, the sum or product will be the same.

EXAMPLE: $3 + 5 = 5 + 3$ In both cases, the sum is 8. This is called the **Commutative Property of Addition.**

$3 \times 5 = 5 \times 3$ In both cases, the product is 15. This is called the **Commutative Property of Multiplication.**

ASSOCIATIVE

If we are adding or multiplying two or more numbers together, we can combine any two at a time, and the final result will be the same.

EXAMPLE:	$(3 + 5) + 8 = 3 + (5 + 8)$
	$8 + 8 = 3 + 13$
	$16 = 16$

In both cases, the sum is 16. This is called the **Associative Property of Addition**. We can "associate" the 5 with either the 3 or the 8.

EXAMPLE:	$(3 \times 5) \times 8 = 3 \times (5 \times 8)$
	$15 \times 8 = 3 \times 40$
	$120 = 120$

In both cases, the product is 120. This is called the **Associative Property of Multiplication**. Again, we can associate the 5 with either the 3 or the 8.

DISTRIBUTIVE

The **Distributive Property** combines both addition and multiplication and introduces a new way of expressing multiplication. When you write a number directly next to a set of parentheses, you multiply whatever is inside the parentheses by that number.

EXAMPLE:	$2(3 + 2 + 5) = 2(10)$
	$= 20$

We can add the numbers in the parentheses together and then multiply the sum by 2.

Or

$$2(3) + 2(2) + 2(5) = 6 + 4 + 10$$
$$= 20$$

We can multiply each number in the parentheses by 2 and then sum the products. Here we are "distributing" the number 2 over each of the numbers in the parentheses.

ORDER OF OPERATIONS

Unless numbers are grouped together with brackets, you always perform multiplications and divisions before additions and subtractions.

EXAMPLE: Simplify the expression: $16 \div 4 + 4 \times 2$

(A) 12
(B) 16
(C) 4
(D) 1

SOLUTION: Remember, we must perform divisions and multiplications first.

$$16 \div 4 \ + \ 4 \times 2 \ =$$
$$4 \ + \ 8 \ = \ 12$$

The correct answer is (A).

Fractions

GLOSSARY

A **proper fraction** is a fraction whose numerator is smaller than its denominator. Thus, a proper fraction has a value less than 1.

EXAMPLE: $\dfrac{5}{6}, \dfrac{7}{9}, \dfrac{2}{3}, \dfrac{17}{18}$ are proper fractions.

An **improper fraction** has a numerator that is larger than or equal to the denominator. Thus, an improper fraction has a value greater than or equal to 1.

EXAMPLE: $\dfrac{6}{5}, \dfrac{9}{7}, \dfrac{5}{3}, \dfrac{4}{4}$ are improper fractions.

A **whole number** can be expressed as an improper fraction by using 1 as the denominator.

EXAMPLE: $8 = \dfrac{8}{1}$

A **mixed number** is a whole number and a fraction. To change a mixed number to an improper fraction, add the numerator of the fraction part to the product of the denominator and the whole number. The result is the numerator. Keep the denominator of the original fraction part.

EXAMPLE: $3\dfrac{1}{4} = \dfrac{3 \cdot 4 + 1}{4} = \dfrac{13}{4}$

Equivalent fractions are fractions that are equal in value. To change any fraction to an equivalent fraction, multiply or divide the numerator and denominator by the same non-zero number.

EXAMPLE: $\dfrac{5}{8} = \dfrac{5 \cdot 3}{8 \cdot 3} = \dfrac{15}{24}$ and $\dfrac{39}{60} = \dfrac{39 \div 3}{60 \div 3} = \dfrac{13}{20}$

Dividing the numerator and denominator by the same number is called **reducing**. To reduce fractions to lowest terms, divide the numerator and denominator by the largest whole number that divides evenly. If you reduce by using a smaller number, be sure to reduce again until you are in lowest terms.

EXAMPLE: $\dfrac{15}{27} = \dfrac{15 \div 3}{27 \div 3} = \dfrac{5}{9}$ and $\dfrac{32}{40} = \dfrac{32 \div 8}{40 \div 8} = \dfrac{4}{5}$ Note here that if you divided by 4 first, you would have had to reduce again by 2: $\dfrac{32}{40} = \dfrac{32 \div 4}{40 \div 4} = \dfrac{8 \div 2}{10 \div 2} = \dfrac{4}{5}$

A **complex fraction** is a fraction whose numerator or denominator, or both, are fractions themselves.

EXAMPLE: $\dfrac{\dfrac{5}{8}}{\dfrac{3}{5}}$ To simplify this complex fraction, divide the fractions.

SOLUTION: $\dfrac{5}{8} \div \dfrac{3}{5} = \dfrac{5}{8} \cdot \dfrac{5}{3} = \dfrac{5 \cdot 5}{8 \cdot 3} = \dfrac{25}{24} = 1\dfrac{1}{24}$

EXAMPLE: 12/48 reduced to lowest terms is equal to

     (A) 1/4
     (B) 6/24
     (C) 4/16
     (D) 3/12

SOLUTION: Each answer correctly reduces the fraction, but only one reduces it to lowest terms. The largest number that divides evenly into 12 and 48 is 12. It goes into 12, 1 time, and into 48, 4 times. The correct answer is (A).

EXAMPLE: 12/72 reduced to lowest terms is equal to

     (A) 6/36
     (B) 1/6
     (C) 4/24
     (D) 3/18

SOLUTION: Again, each answer correctly reduces the fraction, but only one reduces it to lowest terms. The largest number that divides evenly into 12 and 72 is 12. It goes into 12, 1 time, and into 72, 6 times. The correct answer is (B).

EXAMPLE: Which is the largest fraction 7/12, 7/15, or 5/12?

        (A) 7/12
        (B) 7/15
        (C) 5/12
        (D) they are all equal

SOLUTION: If the numerator remains constant, the value of the number becomes smaller as you increase the denominator. So, 7/12 is larger than 7/15. If the denominator remains constant, then the value of the number becomes smaller as you decrease the numerator. So, 7/12 is larger than 5/12. The correct answer is (A).

EXAMPLE: Arrange the numbers, 3 2/3, 17/5, and 3 4/5, in ascending order.

        (A) 3 2/3, 17/5, 3 4/5
        (B) 3 2/3, 3 4/5, 17/5
        (C) 17/5, 3 2/3, 3 4/5
        (D) 3 4/5, 17/5, 3 2/5

SOLUTION: For comparison purposes, it's best to change 17/5 to a mixed number (3 2/5). We know that 3 2/3 is less than 3 4/5, because as we add 1 to both the numerator and denominator, we increase the value of the number. So, we can eliminate (D). 3 2/5 is less than 3 2/3 because the numerators are the same and 5 is greater than 3. So, we can eliminate (A) and (B). The correct answer is (C)

Changing 3 2/3 to an improper fraction (3 2/3 = 11/3) would not have been helpful. It is not immediately obvious which is larger, 17/5 or 11/3.

ADDING FRACTIONS

In order to add fractions, the fractions must have the same denominator.

1. If the denominators are the same, add the numerators, keep the denominator and reduce the fraction, if possible.

    EXAMPLE: $\dfrac{3}{8} + \dfrac{5}{8} = \dfrac{3+5}{8} = \dfrac{8}{8} = 1$

2. If the denominators are not the same, the fractions must be changed to equivalent fractions with the same denominator.

    EXAMPLE: $\dfrac{5}{7} + \dfrac{2}{5} = \dfrac{?}{35} + \dfrac{?}{35} = \dfrac{25}{35} + \dfrac{14}{35} = \dfrac{39}{35} = 1\dfrac{4}{35}$

One way of adding fractions without a common denominator is:

1. Multiply the numerator of the first fraction times the denominator of the second.

2. Multiply the denominator of the first fraction times the numerator of the second.

3. Add the totals in steps 1 and 2. This is the numerator of the sum.

4. Multiply the denominators. This is the denominator.

5. Reduce if possible.

    EXAMPLE: $\dfrac{5}{6} + \dfrac{2}{7} = \dfrac{(5 \cdot 7) + (6 \cdot 2)}{6 \cdot 7} = \dfrac{35 + 12}{42} = \dfrac{47}{42} = 1\dfrac{5}{42}$

SUBTRACTING FRACTIONS

Follow the same procedure as in addition, but subtract the numerators.

EXAMPLE: $\dfrac{5}{8} - \dfrac{1}{8} = \dfrac{5-1}{8} = \dfrac{4}{8} = \dfrac{1}{2}$ and $\dfrac{7}{9} - \dfrac{3}{5} = \dfrac{?}{45} - \dfrac{?}{45} = \dfrac{35}{45} - \dfrac{27}{45} = \dfrac{35-27}{45} = \dfrac{8}{45}$

and $\dfrac{5}{6} - \dfrac{2}{5} = \dfrac{(5 \cdot 5) - (6 \cdot 2)}{6 \cdot 5} = \dfrac{25 - 12}{30} = \dfrac{13}{30}$

ADDING AND SUBTRACTING MIXED NUMBERS

If you are adding or subtracting mixed numbers, you can either:

(1) Change the mixed numbers to improper fractions, then add (subtract).

    OR

(2) Add (subtract) the whole numbers, then add (subtract) the fractions part.

EXAMPLE: $3\dfrac{3}{4} + 5\dfrac{2}{3} = \,?$

SOLUTION: (1) $\quad 3\dfrac{3}{4} + 5\dfrac{2}{3} = \dfrac{15}{4} + \dfrac{17}{3}$

$$= \dfrac{15 \cdot 3 + 4 \cdot 17}{4 \cdot 3}$$

$$= \dfrac{45 + 68}{12} = \dfrac{113}{12}$$

$$= 9\dfrac{5}{12}$$

    OR

(2) $\quad 3\dfrac{3}{4} + 5\dfrac{2}{3} = 3 + 5 = 8 \qquad$ (Add the whole numbers.)

$\dfrac{3}{4} + \dfrac{2}{3} = \dfrac{3 \cdot 3 + 4 \cdot 2}{4 \cdot 3} = \dfrac{9 + 8}{12} = \dfrac{17}{12} = 1\dfrac{5}{12}$ (Add the fractions.)

$8 + 1\dfrac{5}{12} = 9\dfrac{5}{12} \qquad$ (Add them together.)

EXAMPLE: $7\dfrac{3}{8} - 3\dfrac{5}{8}$

SOLUTION: You are not able to subtract 5/8 from 3/8, so you must borrow from the 7.

$$7\dfrac{3}{8} = 6\dfrac{11}{8}$$

$$6\dfrac{11}{8} - 3\dfrac{5}{8} = 3\dfrac{6}{8} = 3\dfrac{3}{4}$$

EXAMPLE: Subtract 7 7/8 from 9 5/12

(A) 2 2/12
(B) 2 2/8
(C) 3 11/24
(D) 1 13/24

SOLUTION: Just from observation, we are subtracting a number that is almost equal to 8 from a number almost equal to 9 1/2. Our answer should be approximately 1 1/2. The only response that comes close to this is (D). However, we'll go through the actual calculation in order to get practice in finding common denominators. You can always find a common denominator by multiplying the 2 denominators together, in this case 8 x 12 = 96. But it's helpful if you can find the least common denominator—in this case, the smallest number that both 8 and 12 divide into evenly. That number is 24. Changing both mixed numbers to fractions of 24 we get:

$$\frac{7}{8} = \frac{?}{24}$$ 8 goes into 24, 3 times; 3 times 7 = 21

$$\frac{7}{8} = \frac{21}{24}$$

$$\frac{5}{12} = \frac{?}{24}$$ 12 goes into 24, 2 times; 2 times 5 = 10

$$\frac{5}{12} = \frac{10}{24}$$

$$9\frac{5}{12} - 7\frac{7}{8} = 9\frac{10}{24} - 7\frac{21}{24}$$

$$= 8\frac{34}{24} - 7\frac{21}{24}$$ Borrow 1 from the 9 and add 24/24 to

$$= 1\frac{13}{24}$$

10/24

The correct answer is (D).

EXAMPLE: Find the sum of $3\frac{2}{3}$ and $2\frac{1}{8}$.

(A) $5\frac{3}{11}$

(B) $6\frac{1}{24}$

(C) $\frac{77}{24}$

(D) $5\frac{19}{24}$

SOLUTION: Just checking relative size, we have a number a little greater than 3 1/2 that we're adding to a number a little larger than 2. We expect a number between 5 1/2 and 6, maybe a little larger. This easily eliminates (A) and (C) as potential answers. We also know that we would need to add 1/3 to 2/3 to reach the next whole number. But 1/8 is less than 1/3, so we know that the sum of 1/8 and 2/3 is less than 1. So, 3 2/3 plus 2 1/8 is less than 6. (D) must be the right answer.

$$3\frac{2}{3} + 2\frac{1}{8} =$$

$$3\frac{16}{24} + 2\frac{3}{24} = 5\frac{19}{24}$$

$$= 5\frac{19}{24}$$

MULTIPLYING FRACTIONS

To multiply fractions, multiply the numerators together, multiply the denominators together, and then reduce, if possible.

EXAMPLE: $\dfrac{3}{5} \cdot \dfrac{2}{9} = \dfrac{3 \cdot 2}{5 \cdot 9} = \dfrac{6}{45} = \dfrac{2}{15}$

Sometimes it is possible to reduce before multiplying. This is called canceling.

EXAMPLE: $\dfrac{5}{8} \times \dfrac{2}{3} = \dfrac{5}{{}_{4}\cancel{8}} \times \dfrac{\cancel{2}^{1}}{3} = \dfrac{5 \times 1}{4 \times 3} = \dfrac{5}{12}$

Remember, after canceling, to multiply the numerators and denominators. To multiply mixed numbers, change to improper fractions first.

EXAMPLE: $3\dfrac{3}{4} \times 1\dfrac{1}{2} = \dfrac{15}{4} \times \dfrac{3}{2} = \dfrac{45}{8} = 5\dfrac{5}{8}$

EXAMPLE: Find the product of 5 1/3 and 2 1/8.

(A) 11 1/3
(B) 272/3
(C) 11 5/8
(D) 9 2/3

SOLUTION: Just checking relative size, we have a number a little greater than 5 multiplying a number a little larger than 2. The product should be a number a little greater than 10. This easily eliminates (B) and (D) as potential answers.

$$5\dfrac{1}{3} = \dfrac{16}{3} \text{ and } 2\dfrac{1}{8} = \dfrac{17}{8}$$

$$\dfrac{{}^{2}\cancel{16}}{3} \times \dfrac{17}{\cancel{8}_{1}} = \dfrac{34}{3} = 11\dfrac{1}{3}$$

The correct answer is (A).

DIVIDING FRACTIONS

To divide fractions, invert the divisor (second fraction), then proceed as in multiplication.

EXAMPLE: $\dfrac{5}{6} \div \dfrac{2}{3} = \dfrac{5}{_2\cancel{6}} \times \dfrac{\cancel{3}^1}{2} = \dfrac{5}{4} = 1\dfrac{1}{4}$

To divide mixed numbers, change the mixed numbers to improper fractions, then divide.

EXAMPLE: $3\dfrac{3}{4} \div 1\dfrac{1}{2} = \dfrac{15}{4} \div \dfrac{3}{2} = \dfrac{{}^5\cancel{15}}{_2\cancel{4}} \times \dfrac{\cancel{2}^1}{\cancel{3}_1} = \dfrac{5}{2} = 2\dfrac{1}{2}$

NOTE: Before changing an improper fraction to a mixed number, be sure to check your answers in the multiple-choice questions to see their form. Some answers may be mixed numbers, and some may be improper fractions.

COMPLEX FRACTIONS

A complex fraction is a fraction whose numerator or denominator (or both) is a fraction.

EXAMPLE: $\dfrac{\frac{5}{6}}{\frac{2}{3}}$ and $\dfrac{\frac{2}{3}}{\left(1+\frac{1}{2}\right)}$ are examples of complex fractions.

To simplify complex fractions, make both the numerator and the denominator proper or improper fractions then divide.

EXAMPLE: $\dfrac{\frac{5}{6}}{\frac{2}{3}}$ is the same as $\dfrac{5}{6} \div \dfrac{2}{3} = \dfrac{5}{6} \times \dfrac{3}{2} = \dfrac{15}{12} = \dfrac{5}{4} = 1\dfrac{1}{4}$

EXAMPLE: $\dfrac{\frac{2}{3}}{1+\frac{1}{2}} = \dfrac{\frac{2}{3}}{\frac{3}{2}} = \dfrac{2}{3} \div \dfrac{3}{2} = \dfrac{2}{3} \times \dfrac{2}{3} = \dfrac{4}{9}$

EXAMPLE: $\dfrac{2-\frac{3}{5}}{1+\frac{1}{3+\frac{1}{4}}} = \dfrac{\frac{10}{5}-\frac{3}{5}}{1+\frac{1}{3\frac{1}{4}}} = \dfrac{\frac{7}{5}}{1+\frac{1}{\frac{13}{4}}} = \dfrac{\frac{7}{5}}{1+\frac{4}{13}} = \dfrac{\frac{7}{5}}{\frac{17}{13}} = \dfrac{7}{5} \div \dfrac{17}{13} = \dfrac{7}{5} \times \dfrac{13}{17} = \dfrac{91}{85}$

15

To simplify expressions such as $1 + \dfrac{1}{\frac{13}{4}}$, remember to treat each complex

fraction as a division problem.

EXAMPLE: Consider $\dfrac{1}{\frac{13}{4}}$ as a division problem: $\dfrac{1}{1} \div \dfrac{13}{4} = \dfrac{1}{1} \times \dfrac{4}{13} = \dfrac{4}{13}$

You might see a shortcut in a problem like this. An expression

such as $1 + \dfrac{3}{\frac{2}{5}}$ can be simplified by inverting the bottom fraction

and multiplying $= 1 + \dfrac{15}{2}$. Prove to yourself that this is the same as

dividing 3 by $\dfrac{2}{5}$.

EXAMPLE: Reduce the complex fraction $\dfrac{\frac{7}{12}}{\frac{21}{8}}$ to lowest terms.

(A) 1 17/32
(B) 8/36
(C) 4 1/2
(D) 2/9

SOLUTION: By just inspecting relative size, you're taking a number that is a little more than half, and dividing by a number more than 2. 1/2 divided by 2 is 1/4. This eliminates (A) and (C) immediately. We can also eliminate (B), because it is not reduced to lowest terms. The answer is (D)

FRACTION PRACTICE PROBLEMS

Try these sample problems:

1. $\dfrac{3}{4} + 2\dfrac{5}{7} =$

2. $\dfrac{8}{9} - \dfrac{2}{5} =$

3. $3\dfrac{2}{3} \times 4\dfrac{1}{2} =$

4. $\dfrac{3}{4} \div 5\dfrac{1}{2} =$

5. $\dfrac{\frac{5}{6}}{\frac{2}{3}} =$

6. Reduce $\dfrac{39}{57}$

7. $17\dfrac{5}{8} \times 128 =$

8. $\dfrac{7}{8} \times \dfrac{2}{3} \div \dfrac{4}{7} =$

9. $12\dfrac{1}{3} - 6\dfrac{3}{4} =$

10. $\dfrac{5}{6} + \dfrac{2}{5} - \dfrac{1}{10} =$

11. $3\dfrac{2}{3} \div 1\dfrac{1}{6} =$

12. $\left(\dfrac{5}{6} + \dfrac{1}{3}\right) \times \left(\dfrac{2}{3} + \dfrac{3}{4}\right) =$

13. $\dfrac{5}{8} \times \dfrac{2}{3} \div \dfrac{5}{6} =$

14. $7\dfrac{1}{2}\left(3\dfrac{1}{4} + 5\dfrac{1}{5}\right) =$

15. $\left(\dfrac{2}{3} \div \dfrac{5}{8}\right) \div 1\dfrac{1}{2} =$

16. $2 + \dfrac{1}{2 + \frac{1}{2}} =$

17. $\dfrac{\frac{5}{6}}{\frac{2}{3} - \frac{1}{6}} =$

ANSWERS:

1. $\dfrac{3}{4} + 2\dfrac{5}{7} = \dfrac{3}{4} + \dfrac{19}{7} = \dfrac{3\cdot 7 + 19\cdot 4}{4\cdot 7} = \dfrac{21 + 76}{28} = \dfrac{97}{28} = 3\dfrac{13}{28}$

2. $\dfrac{8}{9} - \dfrac{2}{5} = \dfrac{8\cdot 5 - 2\cdot 9}{9\cdot 5} = \dfrac{40 - 18}{45} = \dfrac{22}{45}$

3. $3\dfrac{2}{3} \times 4\dfrac{1}{2} = \dfrac{11}{\underset{1}{\cancel{3}}} \times \dfrac{\overset{3}{\cancel{9}}}{2} = \dfrac{33}{2} = 16\dfrac{1}{2}$

4. $\dfrac{3}{4} \div 5\dfrac{1}{2} = \dfrac{3}{4} \div \dfrac{11}{2} = \dfrac{3}{\underset{2}{4}} \times \dfrac{\overset{1}{2}}{11} = \dfrac{3}{22}$

5. $\dfrac{\tfrac{5}{6}}{\tfrac{2}{3}} = \dfrac{5}{6} \div \dfrac{2}{3} = \dfrac{5}{\underset{2}{6}} \times \dfrac{\overset{1}{3}}{2} = \dfrac{5}{4} = 1\dfrac{1}{4}$

6. $\dfrac{39}{57} = \dfrac{39 \div 3}{57 \div 3} = \dfrac{13}{19}$

7. $17\dfrac{5}{8} \times 128 = 17(128) + \dfrac{5}{\underset{1}{8}}(\overset{16}{128}) = 2{,}176 + 80 = 2{,}256$

8. $\dfrac{7}{8} \times \dfrac{2}{3} \div \dfrac{4}{7} = \dfrac{7}{\underset{4}{8}} \times \dfrac{\overset{1}{2}}{3} \times \dfrac{7}{4} = \dfrac{7 \cdot 7}{4 \cdot 3 \cdot 4} = \dfrac{49}{48} = 1\dfrac{1}{48}$

9. $12\dfrac{1}{3} - 6\dfrac{3}{4} = \dfrac{37}{3} - \dfrac{27}{4} = \dfrac{37 \cdot 4 - 27 \cdot 3}{3 \cdot 4} = \dfrac{148 - 81}{12} = \dfrac{67}{12} = 5\dfrac{7}{12}$

10. $\dfrac{5}{6} + \dfrac{2}{5} - \dfrac{1}{10} = \dfrac{25}{30} + \dfrac{12}{30} - \dfrac{3}{30} = \dfrac{25 + 12 - 3}{30} = \dfrac{34}{30} = 1\dfrac{4}{30} = 1\dfrac{2}{15}$

11. $3\dfrac{2}{3} \div 1\dfrac{1}{6} = \dfrac{11}{3} \div \dfrac{7}{6} = \dfrac{11}{\underset{1}{3}} \times \dfrac{\overset{2}{6}}{7} = \dfrac{22}{7} = 3\dfrac{1}{7}$

12. $\left(\dfrac{5}{6} + \dfrac{1}{3}\right) \times \left(\dfrac{2}{3} + \dfrac{3}{4}\right) = \left(\dfrac{5}{6} + \dfrac{2}{6}\right) \times \left(\dfrac{8}{12} + \dfrac{9}{12}\right) = \dfrac{7}{6} \times \dfrac{17}{12} = \dfrac{119}{72} = 1\dfrac{47}{72}$

13. $\dfrac{5}{8} \times \dfrac{2}{3} \div \dfrac{5}{6} = \dfrac{5}{8} \times \dfrac{2}{3} \times \dfrac{6}{5} = \dfrac{\overset{1}{5}}{\underset{4}{8}} \times \dfrac{\overset{1}{2}}{\underset{1}{3}} \times \dfrac{\overset{2}{6}}{\underset{1}{5}} = \dfrac{2}{4} = \dfrac{1}{2}$

14. $7\dfrac{1}{2}\left(3\dfrac{1}{4} + 5\dfrac{1}{5}\right) = 7\dfrac{1}{2}\left(\dfrac{13}{4} + \dfrac{26}{5}\right) = 7\dfrac{1}{2}\left(\dfrac{65}{20} + \dfrac{104}{20}\right) = 7\dfrac{1}{2}\left(\dfrac{169}{20}\right) = \left(\dfrac{15}{2}\right)\left(\dfrac{169}{20}\right) =$

$\dfrac{15 \times 169}{40} = \dfrac{2535}{40} = 63\dfrac{15}{40} = 63\dfrac{3}{8}$

15. $\left(\dfrac{2}{3} \div \dfrac{5}{8}\right) \div 1\dfrac{1}{2} = \dfrac{2}{3} \times \dfrac{8}{5} \times \dfrac{2}{3} = \dfrac{32}{45}$

16. $2 + \dfrac{1}{2 + \dfrac{1}{2}} = 2 + \dfrac{1}{2\dfrac{1}{2}} = 2 + \dfrac{1}{\dfrac{5}{2}} = 2 + \dfrac{2}{5} = 2\dfrac{2}{5}$

17. $\dfrac{\dfrac{5}{6}}{\dfrac{2}{3}-\dfrac{1}{6}} = \dfrac{\dfrac{5}{6}}{\dfrac{4}{6}-\dfrac{1}{6}} = \dfrac{\dfrac{5}{6}}{\dfrac{3}{6}} = \dfrac{\dfrac{5}{6}}{\dfrac{1}{2}} = \dfrac{5}{6} \times \dfrac{2}{1} = \dfrac{10}{6} = 1\dfrac{2}{3}$

Decimals

Decimals are a form of fraction where the denominator is understood to be a power of 10. The first number to the right of the decimal indicates the 10ths place or 10^{-1}, the second the 100ths place or 10^{-2}, the third the 1,000ths place or 10^{-3}, etc.

EXAMPLE: .37 is the equivalent of $\dfrac{37}{100}$

.024 is the equivalent of $\dfrac{24}{1000}$

.8 is the equivalent of $\dfrac{8}{10}$

Note that adding zeros to the right of both the numerator and denominator does not change the value: $.8 = \dfrac{8}{10} = \dfrac{80}{100} = \dfrac{800}{1000}$, etc.

A whole number and a decimal can be written as a mixed number.

EXAMPLE: $38.35 = 38\dfrac{35}{100}$

ADDING AND SUBTRACTING DECIMALS

To add and subtract decimals, line up the decimal points and bring the decimal point straight down. Then add the numbers as if they were whole numbers.

EXAMPLE: 3.5 + 28 + .02 + 5.348 =

$$
\begin{array}{r}
3.500 \\
28.000 \\
.020 \\
+\ 5.348 \\
\hline
36.868
\end{array}
$$

EXAMPLE: 58.236 – 17.04 =

$$
\begin{array}{r}
58.236 \\
-\ 17.040 \\
\hline
41.196
\end{array}
$$

MULTIPLYING DECIMALS

Multiply as if the numbers were whole numbers. Count off from right to left in the answer the total number of decimal places in the factors.

EXAMPLE:

$$
\begin{array}{r}
6.25 \\
\times 3.2 \\
\hline
1.250 \\
18.750 \\
\hline
20.000
\end{array}
\qquad \text{and} \qquad
\begin{array}{r}
.325 \\
\times 1.8 \\
\hline
.2600 \\
.3250 \\
\hline
.5850
\end{array}
$$

DIVIDING DECIMALS

1. Move the decimal out of the divisor.

2. Move the decimal the same number of places in the dividend.

3. Bring the decimal point straight up from the dividend into the quotient.

4. Divide.

NOTE: You may need to add zero after the decimal point is placed in the dividend.

EXAMPLE:

$$
\frac{14}{23} = 23\overline{)14.000} \qquad
\begin{array}{r}
0.608 \\
13.8 \\
\hline
.200 \\
.184 \\
\hline
.016
\end{array}
\qquad \text{and} \qquad
62.50 \div 2.5 = 25\overline{)625.}
\begin{array}{r}
25. \\
50 \\
\hline
125 \\
125 \\
\hline
0
\end{array}
$$

EXAMPLE: 3/8 written as a decimal is equal to

(A) .5
(B) .375
(C) 3.75
(D) .0375

SOLUTION: Changing from fraction to decimal form, you divide the denominator of the fraction into the numerator.

$$\frac{3}{8} = 3 \div 8 = 8\overline{)3.000}$$

$$
\begin{array}{r}
0.375 \\
8\overline{)3.000} \\
\underline{2.4} \\
.600 \\
\underline{.560} \\
.040 \\
\underline{.040} \\
.000
\end{array}
$$

The correct answer is (B).

EXAMPLE: If John did $\frac{1}{4}$ of a job in one day and .3 the next, how much did he complete in both days?

SOLUTION: Either: $\quad \frac{1}{4} + \frac{3}{10} = \frac{5}{20} + \frac{6}{20} = \frac{11}{20} \quad$ or $\quad .25 + .3 = .55$

Look at the answers to see which form to change to.

EXAMPLE: Add $\frac{1}{3}$, .2 and .75.

SOLUTION: Change: .2 to $\frac{2}{10}$ or $\frac{1}{5}$ and $\quad$ .75 to $\frac{75}{100}$ or $\frac{3}{4}$

Add: $\quad \frac{1}{3} + \frac{1}{5} + \frac{3}{4} = \frac{20}{60} + \frac{12}{60} + \frac{45}{60} = \frac{77}{60} = 1\frac{17}{60}$

EXAMPLE: $\frac{2}{5}$ - .068 = .400 - .068 = .332

Can you do this problem by changing .068 to a fraction and proving the answer is the same? You will find this is a little more difficult, but it will allow you to use either form if required.

NOTE: In dealing with fractions such as $\frac{2}{3}, \frac{1}{3}, \frac{1}{9}$ etc., remember they are repeating decimal forms and it is usually best to use the fraction form in calculations and then convert the answer to decimal form if required.

PRACTICE PROBLEMS

Try these sample problems:

1. $.08 \times 7.2 =$

2. $25 + 3.24 + .04 =$

3. Change to fractions:

   A. $.12$

   B. 1.2

   C. $.58$

   D. $.002$

4. Change to decimals:

   A. $\dfrac{5}{8}$

   B. $\dfrac{13}{15}$

   C. $\dfrac{8}{7}$

   D. $3\dfrac{1}{4}$

5. Write 12.42 as a mixed number:

6. $.25\overline{)6.25}$

7. $\dfrac{4}{5} + .72 =$

8. Write $\dfrac{5.6}{100}$ as a decimal:

9. $(2.3 \times 4) + (.025 \times 6.2) =$

10. $\dfrac{1}{5} \times .035 =$

11. $(.3)\left(\dfrac{1}{2}\right) + (.05)\left(\dfrac{4}{5}\right) =$

12. $.27\overline{)1.215}$

13. Add $\dfrac{2}{5}$, .05, and 1.2. Express the answer in decimal form.

14. Add $\frac{2}{5}$, .07, and $\frac{5}{8}$. Express the answer in decimal form.

ANSWERS:

1. .08 × 7.2 = .576

2. 25 + 3.24 + .04 = 25.00
$$\begin{array}{r} 3.24 \\ +\ \ .04 \\ \hline 28.28 \end{array}$$

3. A. $.12 = \frac{12}{100} = \frac{3}{25}$

    B. $1.2 = 1\frac{2}{10} = 1\frac{1}{5} = \frac{6}{5}$

    C. $.58 = \frac{58}{100} = \frac{29}{50}$

    D. $.002 = \frac{2}{1000} = \frac{1}{500}$

4. A. $\frac{5}{8} = 8\overline{)5.000}$ = .625

    B. $\frac{13}{15} = 15\overline{)13.00}$ = .866...

    C. $\frac{8}{7} = 7\overline{)8.00}$ = 1.142$\frac{6}{7}$

    D. $3\frac{1}{4} = \frac{13}{4} = 4\overline{)13.00}$ = 3.25

5. $12.42 = 12\frac{42}{100} = 12\frac{21}{50}$

6.
$$\begin{array}{r} 25. \\ 25\overline{)625} \\ 50 \\ \hline 125 \\ 125 \\ \hline 0 \end{array}$$

7. $\frac{4}{5} + .72$, $5\overline{)4.00}$ = .80, $.8 + .72 = 1.52$

8. $\frac{5.6}{100} = \frac{56}{1000} = .056$

9. (2.3 × 4) + (.025 × 6.2) = 9.2 + .1550 = 9.3550

24

10. $\dfrac{1}{5} \times .035 = \dfrac{.035}{5} = 5\overline{).035}^{\,.007}$

11. $(.3)\left(\dfrac{1}{2}\right) + (.05)\left(\dfrac{4}{5}\right) = \left(\dfrac{3}{10}\right)\left(\dfrac{1}{2}\right) + \left(\dfrac{\overset{1}{\cancel{5}}}{\underset{25}{\cancel{100}}}\right)\left(\dfrac{\overset{1}{\cancel{4}}}{\underset{1}{\cancel{5}}}\right) = \dfrac{3}{20} + \dfrac{1}{25} = \dfrac{15}{100} + \dfrac{4}{100}$

$= \dfrac{19}{100} = .19$ OR $(.3)(.5) + (.05)(.8) = .15 + .04 = .19$

12. $.27\overline{)1.215} = 27\overline{)121.5}^{\,4.5}$

$\phantom{.27\overline{)1.215} = 27\overline{)}}\underline{108}$

$\phantom{.27\overline{)1.215} = 27\overline{)1}}135$

$\phantom{.27\overline{)1.215} = 27\overline{)1}}\underline{135}$

$\phantom{.27\overline{)1.215} = 27\overline{)12}}0$

13. $\dfrac{2}{5} + .05 + 1.2 = .4 + .05 + 1.2 = 1.65$

14. $\dfrac{2}{5} = .4$ and $\dfrac{5}{8} = .625$

Together: .400
.070
.625
1.095

Percents

DECIMALS & PERCENTS

Percent is another way you can refer to a part of 100. 1% is 1 part of 100. 100% is 100 parts of 100, or the entire quantity. The decimal .25 is 25 parts of 100 which is 25%

Changing from decimal form to percent form, shift the decimal point two places to the right and add the percent symbol. Changing from percent to decimal form, shift the decimal point two places to the left and drop the percent symbol.

EXAMPLE: Change .093 to a percent.

SOLUTION: Shift the decimal 2 places to the right = 9.3%.

EXAMPLE: Change 6.4% to a decimal.

SOLUTION: Shift the decimal 2 places to the left, adding a zero in front of the 6 yields .064.

FRACTIONS & PERCENTS

Changing from fraction form to percent form, you have the added step of changing the fraction to a decimal. Then you proceed as before.

Changing from percent to fraction form is a matter of dividing by 100 and dropping the percent symbol.

EXAMPLE: Change 3 1/4% to a fraction

SOLUTION: You can either change 3 1/4 % to 13/4 % or 3.25%

Dividing by 100, we would either add 2 zeros in the denominator or shift the decimal 2 places to the left:

13/4 % = 13/400   3.25% = .0325 = 325/10,000 = 13/400

WORD PROBLEMS

There are 3 elements to a percent problems: the whole, the part (of the whole that you are interested in), and the percentage of the whole that this part represents. The whole is also referred to as the **base**, the part as the **percentage**, and the percent as the **rate**. For example: 3/4 = 75%; 3 is the part, 4 is the whole, and if you have 3 parts of 4, you have 75% of 4.

3/4 = 75%   If we know the part & whole, we divide the whole into the part to find the percent. For example, finding out what part of your total pay is your take home pay.

75% of 4 is 3   If we know the whole & the percent, we multiply the whole by the percent (in decimal form) to find the part. For example, figuring out the tip on a dinner check.

3 is 75% of 4.   If we know the part & the percent, we divide the percent (in decimal form) into the part to find the whole. For example, knowing the sales price of an item and what percent it represents of the original price, and finding out the original price.

Most of the percent problems involve finding one of these 3 parts. When you know which part you are looking for, apply the appropriate rule.

EXAMPLE: Find 30% of $12.50

SOLUTION: "Of" translates to multiplication.
30% of 12.50 = 30% times 12.50 = .3 x 12.5 = 3.75.

A way to simplify this problem is to take 10% of 12.50 by shifting the decimal 1 place to the left -- 10% = $1.25. 30% is 3 times larger or $3.75.

EXAMPLE: What percent of 60 is 24?

SOLUTION: 24 is the part and 60 is the whole. Percent = part/whole
24/60 = 4/10 = .4 = 40%

EXAMPLE: 15% of what number is 5.40?

SOLUTION: The part is 5.40, the percent is 15, we want to find the whole.

Divide the part by the percent:

$$.15\overline{)5.40} = 15.\overline{)540.0}^{\,36.0}$$

EXAMPLE: A Customer left a $4.20 tip. If this represented 15% of the cost of the meal, how much did the meal cost?

(A) $28.00   (B) $6.30   (C) $63.00   (D) $280.00

SOLUTION: The price of the meal is 4.20/.15 = 28. The correct answer is (A).

You should be familiar with some basic formulas involving percent problems.

In interest, the rate is stated as a percent per year. The formula to find the amount of interest is principal times rate times time:

Interest formula: I = P × R × T

EXAMPLE: To find the interest on $1,200 at 6% per year for 6 months:

$$I = 1,200 \times .06 \times \frac{1}{2} = 72 \times \frac{1}{2} = \$36.00$$

NOTE: If a problem expresses the rate in terms of months, then the time must be calculated in terms of months and vice versa.

Be sure you read carefully to see if you are asked to find the interest or the total at the end of the period. In the above problem, the interest would be $36, but the total in the bank would be $1,200 + $36 = $1,236.

To find the amount of principal needed to earn $300 a year interest at 8% for a year, proceed as follows:

$$I = P \times R \times T$$
$$300 = ? \times 8\% \times 1$$
$$300 = ? \times 8\%$$
$$300 \div 8\% = ?$$

$$300 \div \frac{8}{100} = \frac{300}{1} \div \frac{2}{25} = \frac{\overset{150}{\cancel{300}}}{1} \times \frac{25}{\underset{1}{\cancel{2}}} = 150 \times 25 = \$3,750$$

Discount problems are based on the statement:

Amount of Discount = Rate × Original Price

EXAMPLE: Find the rate of discount on an $80 coat, which is discounted to $60.

SOLUTION: Amount of Discount   =   80 - 60 = $20

Amount of Discount   =   Rate × Original Price

$20   =   ? × $80

$$\frac{20}{80} = ?$$

$$\frac{1}{4} = 25\%$$

Remember, if a discount rate is given, the amount paid represents 100% minus the discount rate. For example, on a discount of 30%, the amount paid represents

100% - 30% or 70%.

EXAMPLE: Find the amount paid on a suit selling for $120 if a 30% discount is given.

SOLUTION: 100% - 30% = 70%

70% × 120 = $84.00

A favorite type of percent problem is related to successive discounts. The CLEP examination will test to see if you are aware that a single discount is greater than 2 successive discounts whose sum is the same as the single discount.

EXAMPLE: How much more is a single discount of 40% than 2 successive discounts of 30% and 10%?

SOLUTION: Since a 30% discount means 70% is paid, a 10% discount means 90% is paid, the successive discounts are equivalent to a payment of 90% × 70% or 63%. Therefore, the discount is 100% - 63% = 37%. A single discount of 40% is therefore greater than the successive discounts of 30% and 10%.

Another common type of percent problem involves commissions.

EXAMPLE: A salesman receives a salary of $150 per week, plus a 2% commission on all sales over $300. Find the total salary if his sales were $2,800.

SOLUTION: 2,800 - 300 = 2,500 is the amount on which commission can be taken.

Commission = 2% × 2500 = 50.00

Salary = 150 + 50 = $200

A frequent type of percent problem involves percent increase and decrease.

EXAMPLE: What is the percent increase of a town whose population went from 10,000 to 12,500?

SOLUTION: Percent increase = $\dfrac{\text{amount of increase}}{\text{original amount}}$

$$\% = \dfrac{12,500 - 10,000}{10,000} = \dfrac{2,500}{10,000} = \dfrac{1}{4} = 25\%$$

Remember in percent increase and decrease to use the original number as your denominator, not the new amount.

EXAMPLE: If the attendance at a school football game went down from 2,800 to 2,100, the percent decrease is:

$$\dfrac{2,800 - 2,100}{2,800} = \dfrac{700}{2,800} = \dfrac{1}{4} = 25\%$$

NOTE: The denominator is the original amount, NOT the decreased figure.

Remember in problems with percent amounts greater than 100% of something, you are dealing with an amount greater than the original.

EXAMPLE: If John's salary is 130% of Frank's and John earns $31,200 per year, what is Frank's salary?

SOLUTION: 130% × Frank's salary = 31,200

Frank's salary = 31,200 ÷ 130%
= 31,200 ÷ 1.30
= 24,000

EXAMPLE: If Sue has a coin collection which is valued at $7,200 and Bill's collection is worth 175% of Sue's, how much is Bill's collection worth?

SOLUTION: $7,200 × 175% (Note: This can be done using 1.75 or $1\frac{3}{4}$)

$$\begin{array}{r} 7200 \\ \times\ 1.75 \\ \hline = 12,600 \end{array}$$
  or
$$7,200 \times \frac{7}{4} = 12,600$$

PRACTICE PROBLEMS

Try these problems:

1. Change to fractions:

    A. 28%   B. 3%   C. 5.6%   D. 120%

2. Change to percents:

    A. $\frac{5}{6}$ B. .024   C. $4\frac{1}{5}$ D. .002

3. Change to fractions:

    A. $3\frac{1}{2}$%   B. $\frac{2}{5}$%   C. 15.05%

4. 52% of 720 =

5. 80 ÷ 30% =

6. $2\frac{1}{2}$% of 7,200 =

7. 80% of 90% =

8. 98% ÷ 14% =

9. A.   20% of 80 =

    B.   What % of 10 is 20?

    C.   20% of what number is 14?

10. $\dfrac{1}{2}$ is what % of $\dfrac{2}{3}$?

11. Which of the values below is the largest?

(A) $\dfrac{.1}{.3}$

(B) $\dfrac{.1}{3}$

(C) $\dfrac{1}{.3}$

(D) 310%

12. 42 is $37\dfrac{1}{2}\%$ of?

13. 235% of 780 =

14. 120% of what number is 54?

ANSWERS:

1. A. $28\% = \dfrac{28}{100} = \dfrac{7}{25}$

   B. $3\% = \dfrac{3}{100}$

   C. $5.6\% = \dfrac{5.6}{100} = \dfrac{56}{1000} = \dfrac{7}{125}$

   D. $120\% = \dfrac{120}{100} = \dfrac{6}{5} = 1\dfrac{1}{5}$

2. A. $\dfrac{5}{6} = 6\overline{)5.00}^{\,.83...} = 83\dfrac{1}{3}\%$

   B. $.024 = 2.4\%$

   C. $4\dfrac{1}{5} = 4.2 = 420\%$

   D. $.002 = .2\%$

3. A. $3\dfrac{1}{2}\% = \dfrac{7}{2}\% = \dfrac{7}{200}$

   B. $\dfrac{2}{5}\% = \dfrac{2}{500} = \dfrac{1}{250}$

C. $15.05\% = \dfrac{15.05}{100} = \dfrac{1505}{10000} = \dfrac{301}{2000}$

4. 52% of 720 =

$$\begin{array}{r} 720 \\ \times\,.52 \\ \hline 1440 \\ 3600 \\ \hline 374.40 \end{array}$$

5. $80 \div 30\% = 80 \div \dfrac{3}{10} = 80 \times \dfrac{10}{3} = \dfrac{800}{3} = 266\dfrac{2}{3}$

6. $2\dfrac{1}{2}\%$ of 7,200

$$\begin{array}{r} 7{,}200 \\ \times\,.025 \\ \hline = 36000 \\ 144000 \\ \hline 180.00 \end{array}$$

7. 80% of 90% = .80   or   72%

$$\begin{array}{r} \times\,.90 \\ \hline .7200 \end{array}$$

8. $98\% \div 14\% = 14\overline{)98.00} = 700\%$

$$\begin{array}{r} 7.00 \\ 14\overline{)98.00} \\ \underline{98} \end{array}$$

9. A. 20% of 80 = .20 × 80 = 16.00

   B. What % of 10 is 20? $\dfrac{20}{10} = 2 = 200\%$

   C. 20% of what number is 14?

   $\text{Rate} = \dfrac{\text{Percentage}}{\text{Base}} = \dfrac{14}{20\%} = \dfrac{14}{1} \div \dfrac{1}{5} = \dfrac{14}{1} \times \dfrac{5}{1} = 70$

10. $\dfrac{1}{2}$ is what % of $\dfrac{2}{3}$?

   $\text{Rate} = \dfrac{\text{Percentage}}{\text{Base}} = \dfrac{1}{2} \div \dfrac{2}{3} = \dfrac{1}{2} \times \dfrac{3}{2} = \dfrac{3}{4} = 75\%$

   Proof: $75\% = \dfrac{3}{4}$ $\dfrac{3}{\cancel{4}_2} \times \dfrac{\cancel{2}^1}{3} = \dfrac{3}{6} = \dfrac{1}{2}$

11. The answer is C. Convert all numbers to percent and compare.

    A. $\dfrac{.1}{.3} = 33.3\%$

    B. $\dfrac{.1}{3} = 3.33\%$

    C. $\dfrac{1}{.3} = 333\%$

    D. 310%

12. 42 is $37\dfrac{1}{2}\%$ of what number?

    SOLUTION: $37\dfrac{1}{2}\% = \dfrac{3}{8}$ or $37\dfrac{1}{2}\% = .375$

$$\dfrac{3}{8} \times ? = 42$$

$$? = 42 \div \dfrac{3}{8}$$

$$? = 42 \times \dfrac{8}{3}$$

$$= 112$$

$$\begin{array}{r} 112 \\ 375\overline{)42000} \\ \underline{375} \\ 450 \\ \underline{375} \\ 750 \\ \underline{750} \\ 0 \end{array}$$

13. 235% of 780 = 2.35 × 780 = 1,833.00

14. 120% × ? = 54

    ? = 54 ÷ 120%

    ? = 54 ÷ 1.2

    ? = 45

Table of Fractional and Percent Equivalents

FRACTION	PERCENT	DECIMAL
$\frac{1}{2}$	50%	.5
$\frac{1}{3}$	$33\frac{1}{3}\%$	.33...
$\frac{2}{3}$	$66\frac{2}{3}\%$	.66...
$\frac{1}{4}$	25%	.25
$\frac{3}{4}$	75%	.75
$\frac{1}{5}$	20%	.20
$\frac{2}{5}$	40%	.40
$\frac{3}{5}$	60%	.60
$\frac{4}{5}$	80%	.80
$\frac{1}{6}$	$16\frac{2}{3}\%$	.166...
$\frac{1}{8}$	$12\frac{1}{2}\%$	.125
$\frac{3}{8}$	$37\frac{1}{2}\%$	.375
$\frac{5}{8}$	$62\frac{1}{2}\%$	.625
$\frac{7}{8}$	$87\frac{1}{2}\%$	.875

FRACTION TIPS

■ Any proper fraction with 9 as a denominator repeats the numerator as its decimal form.

EXAMPLE: $\frac{2}{9}$ = .22... $\frac{5}{9}$ = .55...

■ If you can remember the decimal and percent equivalent of $\frac{1}{8}, \frac{1}{6}, \frac{1}{5}$ etc. you can calculate the others by multiplying.

EXAMPLE: If $\frac{1}{8}$ = 12.5%, then $\frac{7}{8}$ = 12.5% × 7 = 87.5%

DIVISION TIPS

■ Even numbers are divisible by 2.

■ A number is divisible by 3 if the sum of the digits is divisible by 3.

EXAMPLE: 714 = 7 + 1 + 4 = 12; 12 is divisible by 3, so 714 is also divisible by 3.

■ A number is divisible by 4 if the number formed by the last two digits is divisible by 4.

EXAMPLE: 748 is divisible by 4 since 48 is divisible by 4.

■ A number is divisible by 5 if the last digit is either 5 or 0.

■ A number is divisible by 6 if it is divisible by 2 and 3.

EXAMPLE: 732 is divisible by 6 since it is even, and the sum of the digits is divisible by 3. 7 + 3 + 2 = 12. 12 is divisible by 3.

■ A number is divisible by 8 if the number formed by the last three digits is divisible by 8.

EXAMPLE: 1,296 is divisible by 8 since 296 is divisible by 8.

■ A number is divisible by 9 if the sum of its digits is divisible by 9.

EXAMPLE: 7,875 is divisible by 9 since 7 + 8 + 7 + 5 = 27 is divisible by 9.

Number System

INTEGER NUMBER RELATIONSHIPS

Integers are whole numbers: 1, 7, 23, 48 and 526 are all examples of integers.

Integers that are divisible by 2 are even integers. Those not divisible by 2 are odd, except for 0. In our list of integers, 48 and 526 are even; 1, 7, and 23 are odd integers. The listing of 1, 7, 23, 48, and 526 represents a **SET** of integers. It is not the set, or listing, of all integers, but it is a set of integers and is represented as:

$$\{1, 7, 23, 48, 526\}$$

48 and 526 represent a **subset** of this set and the subset is represented as {48, 526}.

The set {7, 8, 9, 10} represents a set of **4 consecutive integers**.

The set {12, 14, 16, 18, 20} represents a set of **5 consecutive even integers**.

EXAMPLE: Find the product of 3 consecutive integers whose sum is 30.

    (A) 1000
    (B) 1,320
    (C) 720
    (D) 990

SOLUTION: In a future lesson we'll learn how to solve this problem using algebra. For now, all we need to do is divide 30 (the sum) by 3 (the number of numbers in the sum). We get 10. The numbers will be 10, 1 less than 10, which is 9, and 1 greater than 10, which is 11. Their product is 9 * 11 = 9 * 10 * 11 = 990. The correct answer is (D).

EXAMPLE: If X represents an even number and Y represents an odd number, which of the following are not odd?

    I.   X + Y + 1
    II. 2Y + 3
    III. XY + Y

    (A) I & II
    (B) II & III
    (C) I only
    (D) II only

SOLUTION: In a sum, the only way to get an odd number is if one number is even and the other number is odd. In a product, the only way to get an odd number is if both numbers are odd.

I is even because you have the sum of an odd and an even which is odd. Adding one more makes it even. This eliminates (B) and (D) as possibilities.

II is odd because the product of 2 and Y must be even, and the sum of an even and an odd is odd. This eliminates (A).

We can see that (C) must be the correct answer, but for purposes of discussion, we will evaluate III. III is odd because the product of an odd and an even is even; the sum of an even and an odd is odd. This again would eliminate (B).

If you forget your rules, just substitute an even number for X (2) and an odd number for Y (1) and evaluate each expression:

I. $2 + 1 + 1 = 4$
II. $2 * 1 + 3 = 5$
III. $2 * 1 + 1 = 3$

The only even number is "4".

EXAMPLE: Which of the following must be odd?

I.   The product of two odd numbers

II.   The sum of an odd and an even number

III. The product of two even numbers

(A) I only
(B) II only
(C) I and II only
(D) I and III only

SOLUTION: In examining the 3 possibilities, we find that:

I.   The product of two odd numbers is odd, so I must be included in our solution. We can eliminate choice (B).

II. The sum of an odd and even number is odd, so II must also be included in our solution. We can also eliminate choices (A) and (D). This leaves (C) as the correct answer.

III. Completing our analysis, the product of two even numbers is even, so III is not part of our solution. This once again eliminates (D) as a possible answer.

Therefore, (C) is the correct choice.

Now try the following example.

EXAMPLE: If b is an odd number, which of the following is (are) also odd?

> I.   2b + 1
>
> II. 3(b + 1)
>
> III. 3b + 1

(A) I only
(B) II only
(C) I and II only
(D) I and III only

SOLUTION: In examining the 3 possibilities, I, II, and III, we find that:

> I.   Since b is odd, 2b is even and adding 1 would give us an odd number. Since option I is odd, we can eliminate choice (B).
>
> II. Since b is odd, then b + 1 is even. An odd number times an even number is even. Therefore, II is not correct and we can eliminate choice (C).
>
> III. Since b is odd, three times b must be odd. Adding 1 results in an even number, so III is not correct, and we can eliminate (D).

Therefore, (A) is the correct choice.

This simple chart might help in studying these types of problems:

+ addition	odd	even
odd	= even	= odd
even	= odd	= even

× multiplication	odd	even
odd	= odd	= even
even	= even	= even

If an exponent is involved in an expression, remember that an odd number with any exponent produces an odd number and an even number with any exponent produces an even number. So, if a is odd then $a^3, a^8,$ and a^{10} are also odd. If a is even then $a^2, a^7,$ and a^{12} are also even. You should understand the effect exponents have on positive and negative numbers.

EXAMPLE: If $a = b^2$, what happens to a when b is tripled?

SOLUTION: In this kind of problem you can see the result by substituting 3b for b. Remember to square 3b:

$$a = (3b)^2$$

$$a = 9b^2$$

Therefore, a is nine times larger when b is tripled or, as you might see, $3^2 = 9$. What would happen to a if b were 5 times larger?

A problem to test your understanding of the effect that exponents have on fractions might be as follows:

EXAMPLE: If $P < 0$, what is true of P^3?

        (A) $P^3 > P$

        (B) $P^3 < P$

        (C) $P^3 = P$

        (D) $P^3 = P^2 + P$

        (E)   cannot be determined from given information

SOLUTION: Notice that it does not say that P is a whole number, only that P is negative. If P were a whole number, then raising a negative whole number to the third power would make a smaller or equal negative number.

EXAMPLE: $(-2)3 = -8$.

If P were a negative fraction greater than -1, then raising it to the third power would make a larger negative number. Example: If, $P = -\dfrac{1}{2}$, then $\left(-\dfrac{1}{2}\right)^3 = -\dfrac{1}{8}$ which is greater (closer to zero) than $-\dfrac{1}{2}$. If P were equal to -1, then P^3 would also be -1. We could eliminate choice D immediately, since it violates the laws of exponents. Therefore, none of these choices is correct for all cases, and the correct answer is E.

Some problems might ask you to deal with an algebraic expression and decide if it is odd or even.

EXAMPLE: If a and b are positive odd numbers, is the expression $a^2 + 3ab + b^2$ odd or even?

SOLUTION: Since a is odd, a^2 is odd. Since b is odd, b^2 is odd.

Since a and b are odd, 3ab would be 3 times an odd, or odd. Therefore, we would have the sum of 3 odd numbers which would be odd.

Note: If the expression were $a^2 + 2ab + b^2$ we could recognize it as the square of:

$$(a + b)^2 = a^2 + 2ab + b^2$$

If a and b are odd, then a + b is even and an even number squared is even.

These types of problems are common on the CLEP examination.

SIGNED NUMBERS

Integers can be both positive and negative. Integers include all the numbers that we know as Whole Numbers, but in addition, they also include the opposites of these numbers - the negative numbers.

```
___|___|___|___|___|___|___|___|___|___|___|___|___|
   -6  -5  -4  -3  -2  -1   0   1   2   3   4   5   6
```

Numbers get larger as you move to the right, smaller as you move to the left.

EXAMPLE: -2 > -6   -5 < -3

MULTIPLYING SIGNED NUMBERS

The rules for multiplying signed numbers are:

$$+ \times + = +$$
$$+ \times - = -$$
$$- \times + = -$$
$$- \times - = +$$

If you owe your friend $5.00 and you agree to go double or nothing on the flip of a coin, if you lose, you will double your debt; you will owe $10.00. -5 × 2 = -10. Or, in other words, the product of a negative number and a positive number is negative. We know that in multiplication order is not important to the final result. So,

$$-5 \times 2 = 2 \times (-5) = -10,$$

in other words, the product of a positive number and a negative number is negative.

Finally, if multiplying a negative times a positive yields a negative number, then multiplying a negative by a negative must yield a positive.

In review, if the signs are the same, the product is positive; if the signs are different, the product is negative.

DIVIDING SIGNED NUMBERS

The rules for dividing signed numbers are the same as the rules for multiplying. In fact, in our lesson on fractions, we saw that any division problem can be written as a multiplication problem.

EXAMPLE: Which is larger, (2) × (- 3), (- 2) × (3), or (- 6) × (- 1)

      (A) (2) × (- 3)
      (B) (- 2) × (3)
      (C) (- 6) × (- 1)
      (D) they're all equal

SOLUTION: (A) and (B) will both be negative. (C) is the only positive response, so it's the largest. The correct answer is (C)

ADDING AND SUBTRACTING SIGNED NUMBERS

The rules for combining signed numbers are:

If the signs are the same, add the numbers together and take the common sign.

If the signs are different, take the difference of the 2 numbers and the sign of the larger.

EXAMPLE: -2 + (- 3) =

SOLUTION: To combine 2 signed numbers, first reduce the problem to a single sign between the 2 numbers by applying the same rules you used in multiplication. If the signs are different, the product is negative, so the problem can be rewritten as -2 - 3. The rule for combining these numbers is that if the signs are the same, add the numbers together and take the common sign. In this case the sum of 2 and 3 is 5 and the common sign is negative. $-2 - 3 = -5$.

EXAMPLE: 2 + (- 6) =

SOLUTION: First reduce the problem to a single sign by applying the multiplication rule; the signs are different, the products negative, so the problem reduces to 2 - 6. The rule for combining these numbers is that when the signs are different, take the difference of the 2 numbers and the sign of the larger. In this case, the difference is 4 and the sign in front of the larger number (6) is negative. 2 - 6 = - 4

EXAMPLE: -72 + (-24) =

      (A) 48
      (B) 96
      (C) -96
      (D) -48

SOLUTION: -72 + (-24) = -72 - 24 = -96. The correct answer is (C)

EXAMPLE: -86 - (-95) =

> (A) 9
> (B) 181
> (C) -181
> (D) -9

SOLUTION: -86 - (-95) = -86 + 95 = +9. The correct answer is (A).

ABSOLUTE VALUE

There are times in mathematics when we want to ensure that our answer will be positive. If a number is negative, we'll take its positive counterpart. This is called **Absolute Value**, and is represented by the symbol "| |".

EXAMPLE: Which is larger, |- 24|, (8) (3), or (- 6) (- 4)?

> (A) |- 24|
> (B) (8) (3)
> (C) (- 6) (- 4)
> (D) they're all equal

SOLUTION: In all cases, the value is 24. The answer is (D)

PRIME NUMBERS

A number is a **prime number** if it can only be divided evenly by itself and the number 1. Written as a product, a prime number can only be written as a product of itself and the number 1. For example, 7 can only be written as a product of 7 times 1. 7 is a prime number. 20 can be written as a product of 10 times 2 or 5 times 4 or 20 times 1. 20 is not prime. We call 10 and 2, 5 and 4, factors of 20. The numbers 5, 2, and 2 are the **prime factors** of 20. They are prime numbers that will divide evenly into 20. The factors of 7 are just 1 and 7. A prime number has only two factors, itself and the number 1.

EXAMPLE: How many even prime numbers are there?

> (A) 0
> (B) 1
> (C) 2
> (D) there are an infinite number

SOLUTION: There are many odd prime numbers but there is only one even number that is prime, and that is the number 2. The correct answer is (B)

EXAMPLE: Of the numbers 17, 21, 49, 89, and 105, which are prime?

> (A) 17, 49, & 89
> (B) 49 & 89
> (C) 17 & 89
> (D) 17, 89, 105

SOLUTION: Any number ending in 5 is divisible by 5, so 105 isn't prime. Also, 49 is divisible by 7, so it is not prime. That eliminates (A), (B), and (D) as answers. Obviously the correct answer must be (C), there is no number other than 89 and 1 that will divide evenly into 89 and there is no number other than 17 and 1 that will divide evenly into 17.

EXAMPLE: How many prime numbers are there between 20 and 30?

> (A) 1
> (B) 2
> (C) 3
> (D) 0

SOLUTION: The prime numbers between 20 and 30 are 23 and 29, so there are 2 prime numbers. The correct answer is (B).

EXAMPLE: After the numbers 11 and 13, what is the next set of prime numbers that differ by 2?

> (A) 19 and 21
> (B) 51 and 53
> (C) 27 and 29
> (D) 29 and 31

SOLUTION: Since 21, 27, and 51 are not prime, choices (A), (B), and (C) can be eliminated immediately. The correct answer is (D).

AVERAGES

There are three different types of "average" that we use in mathematics.

The first is the one that is most familiar, the **arithmetic mean**, or just the **mean**. To find the mean of a group of numbers, you add the numbers together and divide by the number of numbers in the group.

EXAMPLE: A 10 question quiz resulted in the following scores. Find the mean test score from the following set of numbers:
{3, 4, 5, 5, 6, 7, 7, 7, 10}

SOLUTION: There are 9 numbers whose sum is 54. The mean is equal to

$$54/9 = 6.$$

The **median** defines a middle point, such that there are an equal number of numbers in our set below this point as there are above this point.

Again, using our example, 6 would also be the median score. There are 4 quiz scores below 6 and 4 above.

The **mode** is the most common occurrence in our set of numbers.

Using our example one more time, we see that the most common quiz score is 7. 7 is the mode.

EXAMPLE: What is the median of set A if A = {2, 8, 10, 10, 14, 15, 18, 19, 21}

    (A) 13
    (B) 14
    (C) 10
    (D) 21

SOLUTION: There are 4 values less than number 14 and four numbers greater than 14, so 14 is the median. The correct answer is (B).

13 = 117 / 9 = the mean; 10 = the mode; 21 is the largest number in the set and obviously can't represent any of the averages.

EXAMPLE: The difference between the mean and the mode of the numbers: 6, 10, 11, 16, 11, 19, and 11 is:

    (A) 11
    (B) 12
    (C) 1
    (D) 0

SOLUTION: The mean of these numbers is equal to the sum, 84, divided by 7, which equals 12. The mode is the most common value, which is 11. The difference between these two numbers is 1. The correct answer is (C). (D) is the difference between the median and the mode.

EXAMPLE: To find the average of (x + 3), x - 6 and 4x - 3:

SOLUTION: Add the group of numbers: (x + 3) + (x - 6) + (4x - 3) = 6x - 6

Then divide by 3: $\frac{6x-6}{3} = 2x - 2$

In some average problems, the numbers may represent different quantities or averages themselves. In this case we have a weighted average.

EXAMPLE: If 8 shirts of one kind cost $10.00 each and 4 of another kind cost $16.00, what is the average cost of all the shirts?

SOLUTION: Here we must multiply to find the total sum first:

8 × 10 = $80.00

4 × 16 = $64.00

Total sum is $80 + $64 = $144.00; 12 shirts cost $144.00. Therefore, the average for all 12 is $12.00.

You can see that if we simply added the two pieces together to get the average, we would end up with:

$$\frac{10+16}{2} = \frac{26}{2} = \$13.00$$

EXAMPLE: If twelve students in a class averaged 70 points out of 100 on an exam, and eighteen students in another class averaged 80 points out of 100 on the same exam, what was the overall average for the whole group?

(A) 77
(B) 75
(C) 73
(D) 76

SOLUTION: Total points scored in the first group equal   70 x 12 = 840

Total points scored in the second group equal   80 x 18 = 1440

Total points scored for both groups equals   = 2280

2,280 / 30 students = 76. The correct answer is (D).

Some problems might give you an average and ask you to find the missing score.

EXAMPLE: If John's scores are 70, 84, 74 and 76, what must he get on his next test to average 80 for the five tests?

SOLUTION: Total sum required = average desired (80) × number of tests (5).

Total points already accumulated = 70 + 84 + 76 + 74 = 304.

Total required less total accumulated: 400 - 304 = 96 must be the score on next test.

EXAMPLE: To find the average of 2x, x + 3 and 3x - 9:

SOLUTION: $\dfrac{2x + (x + 3) + (3x - 9)}{3} = \dfrac{6x - 6}{3} = 2x - 2$

PRACTICE—NUMBER SYSTEM

Try these sample problems:

1. If a is an integer, which of the following is (are) also?

    A. $\dfrac{2a+2}{2}$ B. $\dfrac{a+2}{2}$ C. $\dfrac{2a+1}{2}$

2. If b is odd, what are the 3 odd numbers after b?

3. If a is even, is xyab even or odd where x, y, and b are integers?

4. What happens to c when d is doubled if c = $3d^2$?

5. What is the 7th prime number?

6. A perfect number is any number equal to the SUM of its factors, including 1 but excluding the number itself, what is the first perfect number?

7. If x is odd and y is even, is (xy) + (x + y) odd or even?

8. If P is a negative fraction, which is larger P^2 or P^3?

9. If a is odd and b is even, is $a^3 \times b^3$ odd or even?

10. Is the square of an even number divided by another even number always odd, always even or may it be either?

11. 10 shirts cost \$12 each, 8 shirts of another kind cost \$15 each. What is the average of all 18 shirts?

ANSWERS:

1. A. $\dfrac{2a+2}{2} = \dfrac{2(a+1)}{2} = a+1$ Therefore, A is an integer.

    B. $\dfrac{a+2}{2} = \dfrac{a}{2} + \dfrac{2}{2} = \dfrac{a}{2} + 1$ Therefore, B is an integer only when a is even.

    C. $\dfrac{2a+1}{2} = \dfrac{2a}{2} + \dfrac{1}{2} = a + \dfrac{1}{2}$ Therefore, C is not an integer.

2. b + 2, b + 4, b + 6

3. a is even; therefore, any number of factors will produce an even integer (assuming x, y and b are integers).

4. $$c = 3d^2 \quad c = 3(2d)^2$$
$$c = 3 \times 4d^2$$
$$c = 12d^2$$

Therefore, c is 4 times (or 2^2) larger.

5. The 7th prime number is 17. The first 6 prime numbers are 2, 3, 5, 7, 11, and 13.

6. The first perfect number is 6, since 6 = 3 + 2 + 1. The next perfect number would be 28 because 28 = 14 + 7 + 4 + 2 + 1.

7. Odd, because xy is even, xy + y is odd, and even + odd = odd.

8. P is negative. Therefore, P^3 is negative. P^2 is positive, so $P^2 > P^3$.

9. Even, because a^3 is odd, b^3 is even, and odd × even = even.

10. Since an odd number multiplied by an even produces an even number and so does an even multiplied by an even, it is impossible to determine the quotient of two even numbers.

EXAMPLE: $\dfrac{6^2}{4} = \dfrac{36}{4} = 9(\text{odd}), \quad \dfrac{8^2}{4} = \dfrac{64}{4} = 16(\text{even})$

11. 10 shirts @ $12 = $120

   8 shirts @ $15 = 120

Therefore, 18 shirts = $240

$240 \div 18 = $13.33

Exponents and Radicals

EXPONENTS

In the expression 3^4, 3 is called the base, and 4 is called the exponent. The exponent expresses the number of times that the base should be multiplied by itself. In this case, 3 should be multiplied by itself 4 times, which equals 81.

You need to know certain rules for working with exponents:

MULTIPLYING LIKE BASES

When you multiply like bases, you add the exponents together. For example:

$$2^3 \quad \times \quad 2^2 \quad = \quad 2^5$$

$$(2 \times 2 \times 2) \quad \times \quad (2 \times 2) \quad = 2 \text{ to the fifth power.}$$

This only works when the bases are the same.

$$4^2 \quad \times \quad 4 \quad \times \quad 4^2 \quad = \quad 4^5$$

Remember that 4 is equal to 4 to the first power.

DIVIDING LIKE BASES

EXAMPLE: Divide 3^7 by 3^4.

SOLUTION: $\dfrac{3 \times 3 \times 3 \times 3 \times 3 \times 3 \times 3}{3 \times 3 \times 3 \times 3} = 3^3$

EXAMPLE: Divide 3^4 by 3^7.

SOLUTION: $\dfrac{3 \times 3 \times 3 \times 3}{3 \times 3 \times 3 \times 3 \times 3 \times 3 \times 3} = \dfrac{1}{3^3}$

When we divide like bases, we subtract the smaller exponent from the larger. If the larger exponent is in the denominator, our answer is written in the denominator.

POWERS OF POWERS

EXAMPLE: What is the result of squaring 3 cubed?

SOLUTION: $(3^3)^2 = 3^3 \times 3^3 = 3^6$

We see that the new exponent is equal to the product of the 2 exponents in our problem.

EXAMPLE: Evaluate $(2^6)^2$.

        (A) 2^8
        (B) 2^{12}
        (C) 2^{36}
        (D) 24

SOLUTION: $(2^6)^2 = 2^6 \times 2^6 = 2^{12}$

        The correct answer is (B).

EXAMPLE: Find the product of 2^8 and 4^3.

        (A) 2^{11}
        (B) 2^{14}
        (C) 2^{48}
        (D) 8^{11}

SOLUTION: $2^8 \times (2^2)^3 = 2^8 \times 2^6 = 2^{14}$.

        The correct answer is (B).

RADICALS

SQUARE ROOTS

The opposite of Powers is Roots. Just as you can have powers of 2, (squared), and 3 (cubed), you can have roots of 2, which are called square roots, and roots of 3 or cube roots.

The square root of a number is a number that when multiplied by itself will give you the original number. It is written like this: $\sqrt{}$.

The square root of 49, written as $\sqrt{49}$, is the number when multiplied by itself equals 49. That number is 7. 7 squared is 49.

The square root of 81, or $\sqrt{81}$ is 9.

BASIC LAWS

We already know that $\sqrt{81}$ is 9, because 9 is the number that when multiplied by itself will give you 81. And we know that $\sqrt{9}$ is 3. Consider the following equation:

$$9 = 3 \times 3 = \sqrt{9} \times \sqrt{9} = \sqrt{81}$$

The product of square roots is equal to the square root of the product.

EXAMPLE: $\sqrt{20} \times \sqrt{5} = \sqrt{100} = 10$.

EXAMPLE: $\sqrt{18} \times \sqrt{2} = \sqrt{36} = 6$.

SIMPLIFYING

The $\sqrt{75}$ is not an integer; 75 is not a perfect square, or, in other words, there is no whole number that when multiplied by itself would give you 75. Any number, whether a whole number or fraction, that can't be written without a radical sign, is said to be **irrational**. $\sqrt{75}$ is an **irrational** number. However, even though 75 is not a perfect square, we can simplify $\sqrt{75}$ by saying:

$$\sqrt{75} = \sqrt{25 \times 3} = \sqrt{25} \times \sqrt{3} = 5\sqrt{3}$$

EXAMPLE: Simplify $\sqrt{45}$

      (A) $3\sqrt{5}$
      (B) $9\sqrt{5}$
      (C) $5\sqrt{9}$
      (D) $5\sqrt{20}$

SOLUTION: $\sqrt{45} = \sqrt{9} \times \sqrt{5} = 3 \times \sqrt{5} = 3\sqrt{5}$.

      The correct answer is (A).

EXAMPLE: Which of the following is an irrational number?

      (A) $\sqrt{64}$
      (B) $\dfrac{\sqrt{16}}{9}$
      (C) $\sqrt{19}$
      (D) $\dfrac{3}{\sqrt{25}}$

SOLUTION: (A) is equal to 8; (B) is equal to $\dfrac{4}{9}$; (D) is equal to $\dfrac{3}{5}$. They can all be written without the radical symbol. The correct answer is (B).

EXAMPLE: Simplify:
$$\begin{aligned} \sqrt{48} &= \sqrt{4} \times \sqrt{12} \\ &= \sqrt{4} \times \sqrt{4} \times \sqrt{3} \\ &= 4\sqrt{3} \end{aligned}$$

EXAMPLE: Simplify:$(2\sqrt{3})(4\sqrt{6}) = 2\times 4\times\sqrt{3}\times\sqrt{6}$

$$= 8\times\sqrt{18}$$

$$= 8\times\sqrt{9}\times\sqrt{2}$$

$$= 8\times 3\times\sqrt{2}$$

$$= 24\sqrt{2}$$

EXAMPLE: Combine: $3\sqrt{2}+4\sqrt{2}-\sqrt{2}=(3+4-1)\sqrt{2}=6\sqrt{2}$

EXAMPLE: If $\sqrt{ab}=6$, and a and b are positive integers, which of the following could not be the value of a - b?

A. 5   B. 9   C. 16   D. 4

SOLUTION: Since $\sqrt{ab}=6$, then ab must equal 36. The factors of 36 are 36 × 1, 18 × 2, 12 × 3, 9 × 4, and 6 × 6. By subtracting each set of factors we see that only (D) could not be possible.

RATIONALIZING THE DENOMINATOR

As a practice, we don't leave radicals in the denominator of a fraction. Remember, the denominator names the fraction, and we think of those names as being positive integers: fractions of 2, fractions of 12, etc. We don't speak of fractions of $\sqrt{2}$. So if we see an expression with a radical in the denominator we multiply by a fraction equal to one, a fraction with the radical in both the numerator and the denominator.

EXAMPLE: $\dfrac{1}{\sqrt{2}}=\dfrac{1}{\sqrt{2}}\times\dfrac{\sqrt{2}}{\sqrt{2}}=\dfrac{\sqrt{2}}{\sqrt{2}\sqrt{2}}=\dfrac{\sqrt{2}}{2}$

EXAMPLE: $\dfrac{3}{\sqrt{8}}=$

(A) $\dfrac{3}{4}$

(B) $\dfrac{3\sqrt{2}}{\sqrt{8}}$

(C) $\dfrac{3\sqrt{2}}{4}$

(D) $\dfrac{6}{2\sqrt{2}}$

SOLUTION: Multiplying both the numerator and the denominator by $\sqrt{2}$, we get:

$$\frac{3\sqrt{2}}{\sqrt{8} \times \sqrt{2}}$$

$$\frac{3\sqrt{2}}{\sqrt{16}}$$

$$\frac{3\sqrt{2}}{4}$$

The correct answer is (C).

PRACTICE PROBLEMS -- RADICALS AND EXPONENTS

Try these sample problems:

1. $12^0 \times 10^3 =$

2. $2^{-4} \times 16 =$

3. $x^3 \div x^5 =$

4. $\dfrac{1}{a^{-2}} =$

5. Simplify: $\sqrt{12} + \sqrt{20} =$

6. $(3\sqrt{2})(2\sqrt{6}) =$

7. Rationalize: $\dfrac{1}{2\sqrt{3}}$

8. Simplify: $2\sqrt{3} + \sqrt{75} - \sqrt{27} =$

9. $\sqrt[3]{64} =$

10. Solve for a: $3\sqrt{a-4} = 15$

ANSWERS:

1. $12^0 \times 10^3 = 1 \times 1000 = 1000$

   (Note: any non-zero number to the zero power = 1)

2. $2^{-4} = \dfrac{1}{2^4} = \dfrac{1}{16}, \quad \dfrac{1}{16} \times \dfrac{16}{1} = 1$

3. $x^3 \div x^5 = x^{-2} = \dfrac{1}{x^2}$ Another look: $\dfrac{x^3}{x^5} = \dfrac{x \cdot x \cdot x}{x \cdot x \cdot x \cdot x \cdot x} = \dfrac{1}{x^2}$

4. $\dfrac{1}{a^{-2}} = \dfrac{1}{\frac{1}{a^2}} = \dfrac{1}{1} \div \dfrac{1}{a^2} = \dfrac{1}{1} \times \dfrac{a^2}{1} = a^2$

5. Simplify: $\sqrt{12} + \sqrt{20} = (\sqrt{4} \times \sqrt{3}) + (\sqrt{4} \times \sqrt{5})$
   $$= 2\sqrt{3} + 2\sqrt{5}$$
   $$= 2(\sqrt{3} + \sqrt{5})$$

6. $(3\sqrt{2})(2\sqrt{6}) = 3 \cdot 2 \cdot \sqrt{2} \cdot \sqrt{6}$
   $$= 6 \cdot \sqrt{2} \cdot \sqrt{2} \cdot \sqrt{3}$$
   $$= 6 \cdot 2\sqrt{3}$$
   $$= 12\sqrt{3}$$

7. Rationalize: $\dfrac{1}{2\sqrt{3}} \times \dfrac{\sqrt{3}}{\sqrt{3}} = \dfrac{\sqrt{3}}{2 \cdot 3} = \dfrac{\sqrt{3}}{6}$

8. Simplify: $2\sqrt{3} + \sqrt{75} - \sqrt{27} = 2\sqrt{3} + \sqrt{25} \cdot \sqrt{3} - \sqrt{9} \cdot \sqrt{3}$
   $$= 2\sqrt{3} + 5\sqrt{3} - 3\sqrt{3}$$
   $$= 4\sqrt{3}$$

9. $\sqrt[3]{64} = 4$ (since 4×4×4=64)

10. $3\sqrt{a-4} = 15$

    $\sqrt{a-4} = 5$ dividing by 3

    $a - 4 = 25$ squaring both sides

    $a = 29$

Basic Algebra

ALGEBRAIC EXPRESSIONS

VARIABLES

The expression (5)(4) or 5 * 4 represents the product of 5 and 4 which we know equals 20.

The expression (5)(X) or 5 * X represents the product of 5 and some number. X is a variable. When something is variable, it's changing. In this case, it's the value of X that can change.

$$\text{If } X = 3, \text{ then } 5 * X = 5 * 3 = 15$$

$$\text{If } X = 7, \text{ then } 5 * X = 5 * 7 = 35$$

The product of 5 * X can also be written in its most common form 5X. 5X is called an **Algebraic Term**. Let's look at some more examples of algebraic terms.

3X + 7 is an algebraic expression specifically known as a binomial. Its value is a function of X. We could write this relationship as f(X) = 3X + 7. We will talk more about this format in a later section. 3 is called the **coefficient** of X. 7 is a **constant**

If X = 2, then 3X + 7 = 3(2) + 7 = 6 + 7 = 13

The expression $x^2 + 2x - 15$ is an algebraic expression that can be referred to generally as a **polynomial.** Specifically it is a **trinomial** because it has 3 terms, an x^2 term, an x term and a constant. In this case, the coefficient of the x^2 term is 1, the coefficient of the x term is 2, and the constant is -15.

ADDING & SUBTRACTING ALGEBRAIC EXPRESSIONS

When adding polynomials together remember that you can only add like terms -- terms with the same name. It is the variable that gives the name to the algebraic term.

EXAMPLE: $3x^2 + 7x - 9$

$5x^2 - 12x + 13$

$8x^2 - 5x + 4$

EXAMPLE: $(6x^2 - 7x - 11) - (9x^2 + 4x - 5)$

$6x^2 - 7x - 11 - 9x^2 - 4x + 5 = -3x^2 - 11x - 6$

EXAMPLE: What is the sum of $4x^2 + 5x - 7$ and $x^2 - 3x + 4$?

(A) $4x^2 + 2x - 3$ (B) $5x^2 - 2x + 3$ (C) $5x^2 + 2x - 3$

(D) $5x^2 + 2x + 3$

SOLUTION: $4x^2 + 5x - 7$

$X^2 - 3x + 4$

$\overline{}$

$5x^2 + 2x - 3$

The correct answer is (C)

MULTIPLICATION OF POLYNOMIALS

When you multiply 2 polynomials together, remember that each term of the first polynomial must multiply each term of the second.

EXAMPLE: Consider $(x + 5)(x - 3)$. x must multiply x and -3, and 5 must multiply x and -3.

$x^2 - 3x + 5x - 15 = x^2 + 2x - 15$

EXAMPLE: $(x^2 + 4x - 5)(x - 7) = \quad x^3 - 7x^2 + 4x^2 - 28x - 5x + 35 =$

$x^3 - 3x^2 - 33x + 35$

EXAMPLE: What is the Product of $(3x - 2)$ and $(2x - 7)$?

(A) $6x^2 - 25x - 14$
(B) $6x^2 - 21x - 14$
(C) $6x^2 - 17x + 14$
(D) $6x^2 - 25x + 14$

SOLUTION: The term 3x multiplies both the 2x term and the -7 term. The -2 term multiplies both the 2x term and the -7 term.

$$(3X - 2)(2X - 7) = 6x^2 - 21x - 4x + 14$$

$$= 6x^2 - 25x + 14$$

The correct answer is (D).

FACTORING

Factoring just reverses this process. To write a polynomial in factored form, is to write it as a product of its factors. Let's first take a look at binomials. There are 2 types of factoring problems involving binomials.

COMMON FACTOR

In a Common Factor problem, we are looking for a number and/or a variable that will divide evenly into each algebraic term of the binomial. In any factoring problem, it is the first factor you should look for.

EXAMPLE: Write 6x - 9 in factored form.

SOLUTION: The number that divides evenly into both 6 and 9 is 3. We write 3 outside the parentheses, it's the **common factor**. Inside the parentheses, we write the result of dividing the common factor into each of the binomial terms. 3 goes into 6x, leaving 2x. 3 goes into - 9, leaving - 3.

$$6x - 9 = 3(2x - 3)$$

EXAMPLE: Write, $5x^5 - 25x^3$ in factored form.

SOLUTION: The number that divides evenly into both 5 and 25 is 5. The variable that divides evenly into both x^3 and x^5 is x^3. We write $5x^3$ outside the parentheses, it's the common factor. Inside the parentheses, we write the result of dividing the common factor into each of the binomial terms.

$$5x^5 - 25x^3 = 5x^3 (x^2 - 5)$$

EXAMPLE: The factors of $6x^3 - 9x^9$ are:

(A) $6x^3 (1 - x^6)$
(B) $3x^3 (2 - 3x^6)$
(C) $3x^3 (2 - 3x^3)$
(D) $6x^3 (1 - x^3)$

SOLUTION: The common factor is $3x^3$. Remember, you subtract exponents, not divide. The correct answer is (B).

DIFFERENCE OF 2 PERFECT SQUARES

The second type of binomial factoring problem you might see is the difference of 2 perfect squares.

EXAMPLE: Write $x^2 - 49$ in factor form.

SOLUTION: From what we know about multiplying binomials together, we can get x by multiplying x by x. Similarly, we can get - 49 by multiplying - 7 by + 7. Because the number is the same in both factors, but the signs are different, the x-terms will cancel out, leaving us with just the difference between the square of the first term and the square of the second term.

$$x^2 - 49 = (x + 7)(x - 7)$$

When factoring the difference of two perfect squares, the solution is found by taking the square root of both terms, with one factor being positive and the other negative.

EXAMPLE: Write $x^4 - 81$ in factor form

SOLUTION: $(x^2 + 9)(x^2 - 9) =$

$(x^2 + 9)(x + 3)(x - 3)$

EXAMPLE: The correct factors of $3x^2 - 48$ are

(A) $3(x + 4)(x - 4)$
(B) $3x(x - 16)$
(C) $3(x - 4)(x - 4)$
(D) $(x + 16)(x - 16)$

SOLUTION: This problem combines common factor and the difference of two perfect squares. First, factor out the common factor of 3. What remains inside the parentheses is the difference of two perfect squares.

$3(x^2 - 16) = 3(x + 4)(x - 4)$

The correct answer is (A).

QUADRATIC TRINOMIALS

EXAMPLE: Write $x^2 + 2x - 15$ in factor form.

SOLUTION: If a Quadratic Trinomials can be written as a product of two binomials, then x^2 was the product of the first terms in each binomial, and - 15 was the product of the second terms in each binomial. So, our answer will have the form:

$$(x \quad) (x \quad)$$

The x times x will give us the x^2 term. The x term in the trinomial is the sum of x times some number plus some other number times x. We are looking for 2 numbers whose sum is +2. There are an infinite number of possibilities. Remember, however, that we also know that the product of these numbers is - 15. We know that one number must be positive and the other negative. Finally, we know that the factors of 15 are: 1 & 15, or 3 & 5. Since the sum of the two numbers must be positive 2, the factors we are looking for are + 5 and - 3.

$$(x + 5) (x - 3) = x^2 + 2x - 15$$

EXAMPLE: What are the factors of $x^2 - 3x - 28$?

(A) (x - 4) and (x + 7)
(B) (x + 4) and (x – 7)
(C) (x - 4) and (x – 7)
(D) (x + 4) and (x + 7)

SOLUTION: We are looking for two numbers that add to -3 and multiply to -28. The numbers are +4 and -7. The correct answer is (B).

EXAMPLE: Which is not a factor of $2x^2 - 20x + 48$?

(A) 2
(B) x – 4
(C) x – 6
(D) x + 4

SOLUTION: First, always check for a common factor. In this case, the common factor is 2. We can rewrite the binomial as $2(x^2 - 10x + 24)$.

To factor the trinomial inside the parentheses, we are looking for two numbers whose product is 24 and whose sum is -10. Since the product is positive and the sum is negative, we know that both numbers must be negative. The factors of 24 are 1 & 24, 2 & 12, 3 & 8, and 4 & 6.

- 4 and - 6 are the combination we are looking for. The factored form is:

$$2(x - 4)(x - 6).$$

(A) , (B) , and (C) are factors. The correct answer to the question is (D).

FUNCTIONS

We know what 2 + 5 is equal to, but what about the expression x + 5, where x can be any number? x + 5 = 6, if x = 1; x + 5 = 9, if x = 4; etc.

We say that x + 5 is a Function of x; its value is dependent on the value of x.

We use the notation: f(x) to denote a function of x. In this case, f(x) = x + 5

As we substitute different values for x, we generate new values for f(x). The values that we substitute for x are called the **domain**. The corresponding values of f(x) are called the **range.**

In our section on Modern Math, we will get into a more detailed discussion on sets. For now, it is sufficient to understand that a set is a collection of terms or objects, and the notation for a set is "{ }". For our example, if our domain is the set of even integers between 2 and 10 inclusive (written {2, 4, 6, 8, 10}), then our range is the set {7, 9, 11, 13, 15} or the set of odd integers between 7 and 15 inclusive.

Another way of expressing this relationship is:

D = {x:x = 2, 4, 6, 8, 10}   R = {x:x = 7, 9, 11, 13, 15}. We read this as D (the domain) is equal to all x, where x is equal to 2, 4, 6, 8, or 10. R (the range) is equal to all x where x is equal to 7, 9, 11, 13, or 15.

Again, we will talk more about set theory in a later section.

EXAMPLE: $f(x) = \dfrac{2x + 7}{3x}$, what is f(3)?

SOLUTION: $f(3) = \dfrac{2(3) + 7}{3(3)} = \dfrac{6 + 7}{9} = \dfrac{13}{9} = 1\dfrac{4}{9}$

EXAMPLE: If f(x) = (x² + 12), and g(y) = (y + 1), what is f[g(-3)]?

(A) 21
(B) 16
(C) 3
(D) 28

SOLUTION: Substituting -3 in g(y), we get g(-3) = -3 + 1 = -2. Then substituting -2 in f(x), we get (-2)(-2) + 12 = 4 + 12 = 16. (B) is the correct answer.

EXAMPLE: If f(x) = x² + 2x - 15, then find f(-3)

(A) -12
(B) -27
(C) -30
(D) -15

SOLUTION: f(-3) = (-3)(-3) + 2(-3) - 15 = 9 - 6 - 15 = -12. The correct answer is (A).

EXAMPLE: If f(X) = X² - 7X + 5, and g(Y) = 2Y, then what is f[g(-2)]?

(A) -5
(B) 19
(C) 49
(D) -7

SOLUTION: g(-2) = 2 * (-2) = -4

f(-4) = (-4) * (-4) - 7 * (-4) + 5

16 + 28 + 5 = 49

The correct answer is (C).

PRACTICE PROBLEMS

1. If f(x) = 2x + 3, f(0) =

(A) 0
(B) 2
(C) 3
(D) x

2. If $f(x) = 2x^2 + 1$, then $f(x + 1) =$

   (A) $2x^2 + 2x + 1$
   (B) $2x^2 + 4x + 3$
   (C) $2x^3 + 4x + 2$
   (D) $2x + 3$

3. What is the next value for y in the table below?

x	2	3	4	5
y	6	12	20	?

   (A) 25
   (B) 40
   (C) 30
   (D) 35

4. If $f(x) = x^2$ and $g(y) = 2y$, $f[g(1)] =$

   (A) 4
   (B) 6
   (C) 2
   (D) 8

5. For $y = f(x) = 2x$, $x \in \{1, 2, ...10\}$, the first 3 elements of the range are:

   (A) {1,2,3}
   (B) {2,4,6}
   (C) {∅}
   (D) {8,9,10}

6. If $g(y) = y^2 + 2y + 1$, then $g(y - 1) =$

   (A) $y^2 + 1$
   (B) y^2
   (C) $y^2 + 2$
   (D) $y^2 + 2y$

7. What is the domain of {(1, 2), (2, 0), (3, 5)}?

   (A) {1, 2, 3}
   (B) {0, 2, 5}
   (C) {1, 2}
   (D) {2, 0, 5}

61

ANSWERS:

1. C   F(0)=2(0)+3=3

2. B
$$f(x+1) = 2(x+1)^2 + 1$$
$$= 2(x^2 + 2x + 1) + 1$$
$$= 2x^2 + 4x + 2 + 1$$
$$= 2x^2 + 4x + 3$$

3. C   By squaring each x and adding x to the product you get y. For example: $2^2 + 2 = 6$, $3^2 + 3 = 12$, and $4^2 + 4 = 20$. Therefore, $y = x^2 + x$, so $5^2 + 5 = 30$.

4. A $g(1) = 2(1) = 2$ and $f(2) = 2^2 = 4$.

5. B   The range of a function is the set of y's of the ordered pairs (x, y). For the first three elements of the domain, {1,2,3}, y=2(1)=2, y=2(2)=4, and y=2(3)=6.

6. B
$$g(y-1) = (y-1)^2 + 2(y-1) + 1$$
$$= y^2 - 2y + 1 + 2y - 2 + 1$$
$$= y^2$$

7. A   The domain is the set of first numbers of the ordered pairs. In this set, the first elements of the ordered pairs are 1, 2, and 3.

ALGEBRAIC EQUATIONS

14 + 3 = 17 is an equation. It has terms on both sides of an equal sign and, in fact, the terms must be equal in value.

An equation that has an algebraic term on at least one side of the equal sign is an **algebraic equation**. For example: x + 5 = 50. x is a variable. We are interested in the value of x that will make both sides of the equation equal. What number plus 5 equals 50? In this case, you might be able to figure this out in your head. The number is 45. In general, though, how would we solve an algebraic equation?

Think of an algebraic equation as a balanced scale. If you add something to one side, you must add the same value to the other side. If you multiply one side by a number, you must multiply the other side by the same number. The same goes for subtraction and division. The objective is to have x by itself on one side of the equation. So, starting with:

$$2x + 9 = 15$$
$$\underline{\quad - 9 = -9\quad}$$ we subtract 9 from both sides of the equation.
$$2x \qquad = 6$$

$\dfrac{2x}{2} = \dfrac{6}{2}$ Then we divide both sides by 2. The result is x = 3.

To check, we substitute 3 for x in our original equation, we get 2(3) + 9 which does equal 15.

EXAMPLE: If 3x - 12 = 48, what is x equal to?

        (A) 20
        (B) 12
        (C) 60
        (D) 36

SOLUTION: First we add 12 on both sides of the equation, giving us 3x = 60. Then we divide both sides by 3 giving us x = 20. The correct answer is (A).

EXAMPLE: Find 3 consecutive odd integers whose sum is 87.

SOLUTION: $a + (a + 2) + (a + 4) = 87$

$3a + 6 = 87$

$3a = 81$

$a = 27$

Therefore the three odd numbers are $a = 27$, $a + 2 = 29$, and $a + 4 = 31$. To prove this is correct, sum the numbers: $27 + 29 + 31 = 87$.

EXAMPLE: If the problem said that the sum of the last two of 3 consecutive odd integers is 1 more than three times the first, we would express it as:

SOLUTION: $(x + 2) + (x + 4) = 3(x) + 1$

$2x + 6 = 3x + 1$

$5 = x, x + 2 = 7, x + 4 = 9$

PROOF: $7 + 9 = 1 + 3(5)$

$16 = 16$

EXAMPLE: What is the average of $2x$, $x + 3$, and $3x - 9$?

SOLUTION: Find the sum of the 3 algebraic expressions and divide by 3:

$$\frac{(2x) + (x + 3) + (3x - 9)}{3} = \frac{6x - 6}{3} = 2x - 2$$

EXAMPLE: $3x + 2(x - 4) = 4x - 14 - 2x$

$3x + 2x - 8 = 2x - 14$

$5x - 8 = 2x - 14$

$5x - 2x = 8 - 14$

$3x = -6$

$\dfrac{3x}{3} = \dfrac{-6}{3}$

$x = -2$

PROOF: $3(-2) + 2(-2 - 4) = 4(-2) - 14 - 2(-2)$

$-6 + (-12) = -8 - 14 + 4$

$-18 = -18$

Fractional equations are equations where one or more terms are fractions.

To solve, find the common denominator of all the denominators. Then multiply each term by the common denominator. This will eliminate the fractions and you can solve like a simple equation.

EXAMPLE: $\dfrac{3x}{4} + 1 = \dfrac{x}{3} + 6$

SOLUTION: The common denominator of 4 and 3 is 12.

$$\text{Thus,} \qquad 12\left(\dfrac{3x}{4} + 1\right) = 12\left(\dfrac{x}{3} + 6\right)$$

$$\overset{3}{\cancel{12}} \cdot \dfrac{3x}{\underset{1}{\cancel{4}}} + 12 \cdot 1 = \overset{4}{\cancel{12}} \cdot \dfrac{x}{\underset{1}{\cancel{3}}} + 12 \cdot 6$$

$$9x + 12 = 4x + 72$$

$$5x = 60$$

$$x = 12$$

Decimal equations can be solved by either multiplying by powers of 10 to eliminate the decimals or by combining the decimal parts. To avoid decimal problems, it is usually best to multiply by powers of 10.

EXAMPLE: $.62x + 8.5 = .3x + 8.82$

SOLUTION: Multiply by 100 to remove the decimals:

$$100(.62x) + 100(8.5) = 100(.3x) + 100(8.82)$$

$$62x + 850 = 30x + 882$$

$$32x = 32$$

$$x = 1$$

EXAMPLE: What is the solution set for $| \, 2/3x + 5 \, | = 9$?

      (A) { 6 }
      (B) { 21 }
      (C) { 6, 21 }
      (D) { 6, -21}

SOLUTION: The absolute value will be 9 if the value of the algebraic expression is either +9 or -9. So, there are two equations to solve:

$$\frac{2}{3}x + 5 = 9 \quad \text{and} \quad \frac{2}{3}x + 5 = -9$$

$$\frac{2}{3}x = 4 \qquad\qquad \frac{2}{3}x = -14$$

$$x = 6 \qquad\qquad\qquad x = -21$$

The correct answer is (D).

EXAMPLE: If $\sqrt{x+2} - 3 = 5$, what does x equal?

SOLUTION: To solve this equation, isolate the radical on one side and square the whole equation in order to eliminate the radical symbol.

$$\sqrt{x+2} - 3 = 5$$
$$\sqrt{x+2} - 3 + 3 = 5 + 3 \qquad \text{add 3 to both sides of the equation}$$
$$\sqrt{x+2} = 8$$
$$x + 2 = 64 \qquad \text{square both sides of the equation}$$
$$x + 2 - 2 = 64 - 2 \qquad \text{subtract 2 from both sides of the}$$
$$x = 62 \qquad\qquad\qquad \text{equation}$$

EXAMPLE: If $x^2 + x - 12 = 0$, then x =

(A) -4 or -3
(B) -4 or 3
(C) 4 or -3
(D) 4 or 3

SOLUTION: First, we factor the left side of the equation:

$$(x + 4)(x - 3) = 0$$

If the product is equal to zero, then one of the factors must be zero. So, either:

$$x + 4 = 0 \qquad \text{or} \quad x - 3 = 0$$
$$x = -4 \qquad \text{or} \quad x = 3$$

The correct answer is (B).

EXAMPLE: What is the solution set for the equation x² - 2x = 35.

(A) { 0, 2 }
(B) { 35, 37 }
(C) { 7, - 5 }
(D) { -7, 5 }

SOLUTION: We know nothing about a quadratic sum that is equal to 35, so the first thing we do is subtract 35 from both sides to form a quadratic equation equal to zero. Then we can proceed.

$X^2 - 2x - 35 = 0$

Factoring the left-hand side we get:

$(x - 7) (x + 5) = 0$

$x - 7 = 0$ or $x + 5 = 0$

$x = 7$ or $x = -5$

The correct answer is (C).

QUADRATIC FORMULA

It is possible to solve quadratic equations by using the quadratic formula, but it is usually more time-consuming. If the equation cannot be factored, then you may wish to use the quadratic formula. The formula says the roots of $ax^2 + bx + c = 0$ are:

Quadratic Formula: $x = \dfrac{-b \pm \sqrt{b^2 - 4ac}}{2a}$

where a is the coefficient of the quadratic term, b is the coefficient of the linear term, and c is the constant.

EXAMPLE: Solve for x: x² – 3x = 40.

SOLUTION: Set the equation equal to zero first: x² – 3x - 40 = 0. Therefore, a =1, b = -3, c = -40.

$$x = \frac{-(-3) \pm \sqrt{(-3)^2 - 4(1)(-40)}}{2(1)}$$

$$x = \frac{3 \pm \sqrt{9 + 160}}{2}$$

$$x = \frac{3 \pm \sqrt{169}}{2}$$

$$x = \frac{3 \pm 13}{2}$$

$$x = \frac{3+13}{2} = \frac{16}{2} = 8 \qquad \text{and} \qquad x = \frac{3-13}{2} = \frac{-10}{2} = -5$$

Sometimes the roots (answers) are not whole numbers so you estimate.

EXAMPLE: $x^2 - 7x = 10$

Set = 0: $x^2 - 7x - 10 = 0$

Therefore: $a = 1, b = -7, c = -10$

$$x = \frac{-(-7) \pm \sqrt{(-7)^2 - 4(1)(-10)}}{2(1)}$$

$$x = \frac{7 \pm \sqrt{49 + 40}}{2}$$

$$x = \frac{7 + \sqrt{89}}{2}, \frac{7 - \sqrt{89}}{2} \quad (\sqrt{89} \text{ is approximately } 9.4)$$

$$x = \frac{7 + 9.4}{2}, \frac{7 - 9.4}{2} = 8.2, -1.2 \text{ (approximate)}$$

Word Problems

MOTION PROBLEMS

A common type of word problem is the motion problem. The key to solving motion problems is the formula:

Distance = Rate × Time

OR

D = RT

If you know any two of the three elements, distance, rate or time, you can solve for the other. Many motion problems have two vehicles traveling toward or away from each other. In many cases, the distance, rate or time of the two is the same.

EXAMPLE: Two cars are traveling toward each other at 50 MPH and 40 MPH respectively. If they leave at the same time from a distance of 270 miles apart, when will they meet?

SOLUTION: In this problem, the cars travel the same amount of time.

Since D = R × T, then $T = \dfrac{D}{R}$. The first car's time $= \dfrac{x \text{ miles}}{50 \text{ MPH}}$, and the second car's time $= \dfrac{270 - x}{40 \text{ MPH}}$. (Note: 270 - x is the distance remaining). Therefore, since the time traveled is the same: $\dfrac{x}{50} = \dfrac{270 - x}{40}$.

Cross multiplying: $40x = 50(270 - x)$

Dividing by 10:

$$4x = 5(270 - x)$$
$$4x = 1350 - 5x$$
$$9x = 1350$$
$$x = 150 \text{ miles}$$

The first car travels a distance of 150 miles at 50 MPH. Therefore, the time will be $\dfrac{150}{50} = 3$ hours.

Another way to solve this problem is to figure out how fast the cars are traveling toward each other: 40 + 50 MPH. Divide this result into the total distance traveled: $\dfrac{270}{90} = 3 \text{ hours}$.

A useful ratio in motion problems is 30 MPH = 44' per second. This ratio is useful with problems that require you to change MPH to feet per second.

EXAMPLE: A plane travels at 600 MPH. How many feet per second is this?

SOLUTION: $\dfrac{30 \text{ MPH}}{44 \text{ FPS}} = \dfrac{600 \text{ MPH}}{x \text{ FPS}}$

$$30x = 600 \times 44$$
$$x = 20 \times 44$$
$$x = 880 \text{ feet per second}$$

WORK PROBLEMS

Work problems are often found on the exam. The key to work problems is:

1. Express the rate of work done each day (or hour) as a fraction.

2. Let **x** represent the time required for each of two or more working together to complete the job.

3. Let the completed job be represented by 1.

EXAMPLE: If John does a job in 6 hours, Bob does the same job in 3 hours; how long will it take both, working together?

SOLUTION: John does $\dfrac{1}{6}x$ per hour. Bob does $\dfrac{1}{3}x$ per hour.

Together: $\dfrac{1}{6}x + \dfrac{1}{3}x = 1$

$$6 \cdot \dfrac{1}{6}x + 6 \cdot \dfrac{1}{3}x = 6 \cdot 1$$
$$1x + 2x = 6$$
$$3x = 6$$
$$x = 2 \text{ hours}$$

CONSECUTIVE INTEGER PROBLEMS

Consecutive integer problems can only be solved when the integers are represented correctly.

EXAMPLE: If the sum of the last two of 3 consecutive odd integers is 5 less than 3 times the first, what are the integers?

SOLUTION: x is the first, x + 2 is the second, x + 4 is the third odd integer.

Therefore, $(x + 2) + (x + 4) = 3x - 5$
solving, $\qquad 2x + 6 = 3x - 5$
$$11 = x$$
Therefore the integers are x = 11, x + 2 = 13, and x + 4 = 15.

DIGIT PROBLEMS

Digit problems are not difficult if you remember that a two-digit number can be represented by $10x + y$ where x represents the tens digit and y represents the units digit. A three-digit number would be represented by $100x + 10y + z$. To represent the two-digit number with the digit reversed, write $10y + x$.

EXAMPLE: A two-digit number is 36 more than the number with the digits reversed. If the tens digit is twice the unit's digit, what is the number?

SOLUTION: x = Ten's Digit, y = Unit's Digit

1. $x = 2y$ (tens digit is twice the unit's digit)

2. Digit $= 10x + y$

3. Digit reversed $= 10y + x$

Therefore, $10x + y = 36 + 10y + x$

Like terms on same side: $9x - 9y = 36$

Divide by 9: $x - y = 4$

Substitute equation 1: $2y - y = 4$

$$y = 4$$

Solve for x: $x = 2(y) = 2(4) = 8$

The number is 84, which is 36 more than 48.

FRACTION PROBLEMS

To solve fraction problems, represent the numerator or denominator with an unknown. Multiply the equation by the common denominator of the denominators.

EXAMPLE: A fraction equal to $\frac{3}{5}$ has the numerator increased by 4 and denominator doubled. The new fraction is $\frac{1}{2}$. Find the original fraction.

SOLUTION: Since the original fraction is equal to $\dfrac{3}{5}$, we can represent it by

$\dfrac{3x}{5x}$.

Therefore: $\dfrac{3x + 4}{2(5x)} = \dfrac{1}{2}$

$$2(3x + 4) = 10x$$

$$6x + 8 = 10x$$

$$8 = 4x$$

$$2 = x$$

Since x = 2: $\dfrac{3x}{5x} = \dfrac{3(2)}{5(2)} = \dfrac{6}{10}$ was the original fraction.

AGE PROBLEMS

Age problems test your ability to represent a person's age in the future or in the past.

EXAMPLE: To represent a person's age 5 years from now, if he was 25 years old x number of years ago, write:

SOLUTION:
$25 + x$ = present age
$(25 + x) + 5$ = age in 5 years
$30 + x$ = age in 5 years

PRACTICE — WORD PROBLEMS

1. A merchant sells two blends of coffee. In the first, 2 pounds of $0.60 coffee are mixed with one pound of $0.90 coffee. In the second, 2 pounds of $0.90 coffee are mixed with one pound of $0.60 coffee. What is his additional profit if he sells 30 pounds of the first blend at the average price per pound of the second blend?

2. A man drives 100 miles one way on a trip at an average speed of 30 MPH. His return rate was only an average of 20 MPH. What is his average rate for the entire trip?

3. A pool has 2 inlet pipes. Each pipe used alone can fill the entire pool in 12 hours. What part of the pool can be filled in 3 hours if both pipes are used together?

4. John has a coin collection of 20 coins. The number of half-dollars is 2 more than the number of quarters, and the number of dimes is 6 more than the number of quarters. What is the value of the collection?

72

5. A missile goes 5,400 miles in 45 minutes. How much farther will it go in another 50 minutes at the same speed?

6. If John takes 30 minutes to get to work 20 miles away, how much faster must he travel if he leaves 10 minutes late?

7. How much money is invested at 4% and 5% if:

   a. the total is $7500?

   b. the interest on the amount invested at 4% is $75 more than the amount at 5%?

8. Find two numbers such that the larger is 3 times the smaller, and the sum of the larger and twice the smaller is 45.

9. Find four consecutive odd integers such that the sum of the first 3 integers is six more than the last.

10. In a football league each team plays every other team. If there are 28 games played, how many teams are in the league?

ANSWERS:

1. In the first blend the value is 2(.60) + .90 = $2.10. The second is worth 2(.90) + 60 = $2.40. The average price of the first is $\frac{2.10}{3}$ = $0.70 per lb. The average price of the second is $\frac{2.40}{3}$ = $0.80 per lb. If the first mixture is sold at the average price of the second, he will make $0.10 per lb. extra. For 30 pounds, therefore, he will make 30 · (.10) = $3.00 profit.

2. Apply the following formula to solve this problem:

   $$\text{Rate (for entire trip)} = \frac{2(\text{rate one way})(\text{rate other way})}{\text{rate one way} + \text{rate other way}}$$

   Substituting you get: $\frac{2(30)(20)}{30 + 20} = \frac{1200}{50} = 24$

   Note that in this problem the distance traveled is immaterial because the rate given for each way is an average. You can, however, solve this problem using the distance given and the formula Distance = Rate × Time.

3. Letting the amount done by each pipe be represented by a fraction can solve this problem.

Therefore: $\dfrac{x}{12} + \dfrac{x}{12} = 1$

$$x + x = 12$$
$$2x = 12$$
$$x = 6$$

Therefore both pipes together can fill the pool in 6 hours. In 3 hours $\dfrac{3}{6} = \dfrac{1}{2}$ of the pool will be filled.

4. Let x represent the number of quarters, x + 2 represent the number of half dollars, and x + 6 represent the number of dimes.

Therefore: $x + (x + 2) + (x + 6) = 20$

$$3x + 8 = 20$$
$$3x = 12$$
$$x = 4$$

and 4(.25) + 6(.50) + 10(.10) = $1.00 + $3.00 + $1.00 = $5.00

5. $$D = R \times T$$
$$5{,}400 = R \times 45 \text{ min.}$$
$$R = 120 \text{ mi./min.}$$

Therefore: $D = 50$ min. (120 miles/min.)
$$D = 6{,}000 \text{ miles}$$

You can also solve by using ratio:

$\dfrac{5{,}400}{R} = \dfrac{45}{50}$.

6. In order to go 20 miles in 20 min. (10 min. late) John will have to travel 60 MPH (a mile a minute). His usual rate is $R = \dfrac{20}{.5} = 40$ MPH. Therefore he will have to go 20 MPH faster.

7. Let x represent the amount invested at 4%. Then (7500 - x) is the amount at 5%. Therefore: $4\%(x) = \$75 + 5\%(7500-x)$

$$4\%x = \$75 + 375 - 5\%x$$

$$9\%x = 450$$

$$x = \frac{450}{.09} = \$5000 \text{ at } 4\%$$

$$7500 - 5000 = 2500 \text{ at } 5\%$$

Can you prove this answer?

8. Let x = smaller number and 3x = larger number.

$$3x + 2(x) = 45$$

$$5x = 45$$

$$x = 9$$

$$3x = 27$$

9. x = 1st, x+ 2 = 2nd, x + 4 = 3rd, x + 6 = 4th.

$$x + (x + 2) + (x + 4) = 6 + (x + 6)$$

$$3x + 6 = 6 + x + 6$$

$$2x = 6$$

$$x = 3$$

Therefore the answer is: 3, 5, 7, 9.

10. If each team plays every other team, let n = number of teams. Then n(n - 1) = total number of games (since no team plays itself). However, this total is twice as large since "A plays B" is the same as "B plays A".

Therefore: $\dfrac{n(n-1)}{2} = 28$

$$n(n - 1) = 56$$

$$n^2 - n - 56 = 0$$

Factor: $(n - 8)(n + 7) = 0$

$$n - 8 = 0 \qquad n + 7 = 0$$

$$n = 8 \qquad n = \text{-7 (reject, can't be negative)}$$

SIMULTANEOUS SYSTEMS

When there are two unknowns with exponents of 1 in an equation, it is necessary to have two equations to solve for both variables.

EXAMPLE: If $3x + y = 18$ and $x - y = 2$, we can solve for x and y by combining the terms to eliminate one of the variables.

Thus, $3x + y = 18$ Proof:

$ + x - y = 2$ $3(5) + 3 = 18$ True

$ 4x = 20$ $5 - 3 = 2$ True

$ x = 5$ Therefore the solutions are both correct.

If $x = 5$, then $5 - y = 2$ and $y = 3$.

Since the equations are statements of equality, it is possible to multiply the equations so that the coefficients for one of the variables are the same.

EXAMPLE: $3x + 2y = 10$ and $2x - 3y = 11$

SOLUTION: Multiply the first equation by 3, the second by 2, then combine:

$$9x + 6y = 30$$
$$4x - 6y = 22$$
$$13x = 52$$
$$x = 4$$
$$3(4) + 2y = 10 \quad \text{Find y by substituting 4 for x.}$$
$$12 + 2y = 10$$
$$2y = -2$$
$$y = -1$$

PROOF: $3(4) + 2(-1) = 10$ True

$ 2(4) - 3(-1) = 11$ True

Systems of equations are most often found in word problems where two variables are found. If you express the problem with two equations using the two variables, you can solve the problem with the simultaneous equations.

EXAMPLE: A man paid $39.00 for 5 shirts and 4 hats. At the same price, 4 shirts and 5 hats cost $42.00. What is the price of each?

SOLUTION: Let x = price of hats and y = price of shirts.

Therefore: $4x + 5y = 39$, and

$5x + 4y = 42$

Multiply each to get a common coefficient for x:

$5(4x) + 5(5y) = 5(39)$

$4(5x) + 4(4y) = 4(42)$

Subtracting:$20x + 25y = 195$ (which eliminates x)

$- 20x + 16y = 168$

$9y = 27$

$y = 3$

Therefore, the shirts cost \$3.00 each. To find the price of hats, substitute:

$4x + 5(3) = 39$

$4x + 15 = 39$

$4x = 24$

$x = 6$

Therefore, the hats cost \$6.00 each.

PRACTICE — BASIC ALGEBRA

Try these sample problems:

1. $3a - 5a + 7a =$

2. $(5a^2)(-2a^3) =$

3. $4c - 3a + 6c + a =$

4. $\dfrac{2a^2}{3b} \cdot \dfrac{18b^4}{6a^3} =$

5. Solve:

A. $\dfrac{3ab}{2c} \div \dfrac{12b^2}{10c^2} =$

B. $\left(\dfrac{2}{3}\right)^2 \cdot \left(\dfrac{1}{2}\right)^2 =$

6. Evaluate $3a^2 - 5a - 4$ if $a = -2$

7. Factor:

    A. $a^2 - 7a - 18$

    B. $3a^2 - 11a + 6$

8. Simplify:

    A. $(3x - 5)(2x + 4)$

    B. $3(x + y) - (3x - y)$

9. Solve for x:
    A. $3x - 7 = 2(x - 5)$

    B. $3(x - y) = 2ab$

10. Solve for x:

    $x = y + 2$
    $2x - 5y = 1$

11. Solve for a:

    $a + b = 4$
    $a - b = c$

12. Solve for x:

    $x^2 - 4x = 21$

13. Add: $\dfrac{3a}{x} + \dfrac{4}{y} =$

14. Solve for x: $\dfrac{3}{7} = \dfrac{12}{x}$

15. Solve for x: $3.5x + .7 = 7.7$

16. Solve for y: $y^2 - y = 12$

17. $|3a - 1| = 11$

18. Solve using the quadratic formula:

    $y^2 - 3y = 5$

ANSWERS:

1. $3a - 5a + 7a = 5a$

2. $(5a^2)(-2a^3) = -5 \times -2 \times a^2 \times a^3 = -10a^5$

3. $4c - 3a + 6c + a = 10c - 2a$

4. $\dfrac{\overset{1}{\cancel{2a^2}}}{\underset{1}{\cancel{3b}}} \cdot \dfrac{\overset{6b^3}{\cancel{18b^4}}}{\underset{3a}{\cancel{6a^3}}} = \dfrac{1}{1} \cdot \dfrac{6b^3}{3a} = \dfrac{6b^3}{3a} = \dfrac{2b^3}{a}$

78

5. A. $\dfrac{3ab}{2c} \div \dfrac{12b^2}{10c^2} = \dfrac{\overset{1a}{\cancel{3ab}}}{1} \cdot \dfrac{\overset{5c}{\cancel{10c^2}}}{\underset{4b}{\cancel{12b^2}}} = \dfrac{1a}{1} \cdot \dfrac{5c}{4b} = \dfrac{5ac}{4b}$

   B. $\left(\dfrac{2}{3}\right)^2 \cdot \left(\dfrac{1}{2}\right)^2 = \dfrac{\overset{1}{\cancel{4}}}{9} \cdot \dfrac{1}{\underset{1}{\cancel{4}}} = \dfrac{1}{9}$

6. $3a^2 - 5a - 4 = 3(-2)^2 - 5(-2) - 4$

   $\qquad\qquad\quad = 3(4) + 10 - 4$

   $\qquad\qquad\quad = 12 + 10 - 4$

   $\qquad\qquad\quad = 18$

7. A. $a^2 - 7a - 18 = (a - \quad)(a + \quad)$

   $\qquad\qquad\qquad\quad = (a - 9)(a + 2)$

   B. $3a^2 - 11a + 6 = (3a - 2)(a - 3)$

8. A. $(3x - 5)(2x + 4) =$

   $\qquad\qquad\quad = (3x)(2x) + (3x)(4) + (-5)(2x) + (-5)(4)$

   $\qquad\qquad\quad = 6x^2 + 12x - 10x - 20$

   $\qquad\qquad\quad = 6x^2 + 2x - 20$

   B. $3(x + y) - (3x - y) = 3x + 3y - 3x + y = 4y$

9. A. $3x - 7 = 2(x - 5)$ Proof: $3(-3) - 7 = 2(-3 - 5)$

   $\qquad 3x - 7 = 2x - 10$ $\qquad\qquad -9 - 7 = 2(-8)$

   $\qquad\qquad\ x = -3$ $\qquad\qquad\qquad -16 = -16$

   B. $3(x - y) = 2ab$ or $3(x - y) = 2ab$

   $\qquad x - y = \dfrac{2ab}{3}$ $\qquad 3x - 3y = 2ab$

   $\qquad\quad x = \dfrac{2ab}{3} + y$ $\qquad\quad 3x = 2ab + 3y$

   $\qquad\quad x = \dfrac{2ab + 3y}{3}$ $\qquad\quad\ x = \dfrac{2ab + 3y}{3}$

10. $x = y + 2$ and $2x - 5y = 1$, by substitution:

$$2(y + 2) - 5y = 1$$
$$2y + 4 - 5y = 1$$
$$-3y + 4 = 1$$
$$-3y = -3$$
$$y = 1$$

Therefore: $x = y + 2$
$$x = 1 + 2$$
$$x = 3$$

11. $$a + b = 4$$
$$\underline{a - b = c}$$
$$2a = 4 + c$$
$$a = \frac{4 + c}{2}$$

12. $x^2 - 4x = 21$ change to: $x^2 - 4x - 21 = 0$
$$(x - 7)(x + 3) = 0$$
$$x - 7 = 0 \quad x + 3 = 0$$
$$x = 7 \qquad x = -3$$

13. $$\frac{3a}{x} + \frac{4}{y} = \frac{3a \cdot y + 4 \cdot x}{xy} = \frac{3ay + 4x}{xy}$$

14. $$\frac{3}{7} = \frac{12}{x}$$
$$3x = 84$$
$$x = \frac{84}{3} = 28$$

15. $3.5x + .7 = 7.7$ Proof: $3.5(2) + .7 = 7.7$
$$35x + 7 = 77 \qquad\qquad 7 + .7 = 7.7$$
$$35x = 70 \qquad\qquad\quad 7.7 = 7.7$$
$$x = 2$$

16. $y^2 - y = 12$ Proof: $4^2 - 4 = 12$ $(-3)^2 - (-3) = 12$

   1. $y^2 - y - 12 = 0$ $16 - 4 = 12$ $9 + 3 = 12$

   2. $(y - 4)(y + 3) = 0$ $12 = 12$ $12 = 12$

   3. $y - 4 = 0$ $y + 3 = 0$

   4. $y = 4$ $y = -3$

17. $+(3a - 1) = 11$ $-(3a - 1) = 11$

   $3a - 1 = 11$ $-3a + 1 = 11$

    $3a = 12$ $-3a = 10$

     $a = 4$ $a = -\dfrac{10}{3} = -3\dfrac{1}{3}$

18. $y^2 - 3y - 5 = 0$

   $a = 1, \ b = -3, \ c = -5$

$$x = \frac{-(-3) \pm \sqrt{(-3)^2 - 4(1)(-5)}}{2(1)}$$

$$x = \frac{3 \pm \sqrt{9 + 20}}{2}$$

$$x = \frac{3 + \sqrt{29}}{2}, \frac{3 - \sqrt{29}}{2}$$

$$x = \frac{3 + 5.38}{2}, \frac{3 - 5.38}{2}$$

$$x = 4.19, \quad x = -1.19$$

SOLVING INEQUALITIES

An inequality is a statement that two quantities are not equal. One may be greater than or less than another. The symbols of inequality are:

$a > b$ (read a is greater than b)

$a < b$ (read a is less than b)

$a \geq b$ (read a is greater than or equal to b)

$a \leq b$ (read a is less than or equal to b)

$a \neq b$ (read a is not equal to b)

Both sides of an inequality can be added to without changing the direction of the inequality.

EXAMPLE: If $a > b$, then $a + c > b + c$

Remember, this principle includes subtraction also, since subtraction is defined as adding negatives.

Both sides of an inequality can be multiplied and divided by the same positive number without changing the direction of the inequality.

EXAMPLE: If $a > b$ and c is positive $(c > 0)$, then $a \cdot c > b \cdot c$ and $\dfrac{a}{c} > \dfrac{b}{c}$.

However, if c is negative $(c < 0)$, then $ac < bc$ and $\dfrac{a}{c} < \dfrac{b}{c}$.

We can deal with inequalities the same way we solve equations.

EXAMPLE: If $3x + 2 > x + 10$ then,

$\qquad 2x > 8$ $\qquad$ (subtracting x and 2 from both sides)

$\qquad x > 4$ $\qquad$ (dividing by 2)

Notice that these are the same steps we would use if we were solving an equation.

Our answer says that $3x + 2$ is greater than $x + 10$ whenever x is greater than 4. (Try substituting a number greater than four.)

Remember, if an inequality has a negative coefficient, reverse the order (direction) in solving.

EXAMPLE: If $\qquad -5x - 2 > -x + 18$

then $\qquad -4x > 20$

(dividing by -4) $\quad x < -5$

This says that $-5x - 2$ is greater than $-x + 18$ whenever x is less than -5. (Try substituting a number less than -5 to prove).

Some inequalities are presented as representing a range of numbers.

EXAMPLE: $4 < a < 10$ says a is greater than 4, but less than 10.

You can deal with one or both ends of the inequality.

EXAMPLE: If $3 < r < 9$ and $1 < s < 10$, then the

range: $(3 + 1) < r + s < (9 + 10)$

and $(3 - 10) < r - s < (9 - 1)$

$(3 \cdot 1) < r \cdot s < (9 \cdot 10)$

$(3 \div 10) < r \div s < (9 \div 1)$

If an inequality is presented and you are asked to compare parts, use one of the principles involving adding or multiplying.

EXAMPLE: If y is positive and $\dfrac{x}{y} > 3$, which is greater, x or 2y?

SOLUTION: If we multiply both sides of the inequality by y we get:

$$y \cdot \dfrac{x}{y} > 3$$
$$x > 3y$$

Therefore, x is greater than 2y.

PRACTICE — INEQUALITIES

Try these sample problems:

1. If $a > b$, are integers then which of the following is/are not necessarily true?

    A. $ac > bc$ B. $a^2 > b^2$ C. $a + c > b + c$ D. $\dfrac{a}{c} > \dfrac{b}{c}$

2. Solve for x: $3x + 2 > 11$

3. Solve for x: $-3x + 2 > 11$

4. If $\dfrac{P}{N} > 2$, then $P > ?$ (N is not zero.)

5. $4 < x < 10$ and $-2 < y < 12$. What is the range of $x + y$?

6. Which of the following could not be true if $a + a^2 < 1$?

    A. $a > 1$ B. $a > -1$ C. $a < 0$ D. $a > 0$

7. Solve for x: $3x - 8 < x - 18$

8. If $0 < a - b < 3$, then a ? b

9. If $a > 0$, which is greater, a^2 or a?

10. If $a < 0$, which is greater, a^2 or a?

11. If $a < 5 < b$ (a and b are positive) then which is true:

    A. $\dfrac{1}{a} < \dfrac{1}{5} < \dfrac{1}{b}$ B. $\dfrac{1}{b} < \dfrac{1}{5} < \dfrac{1}{a}$ C. $\dfrac{1}{a} < \dfrac{1}{b} < \dfrac{1}{5}$ D. $\dfrac{1}{b} < \dfrac{1}{a} < \dfrac{1}{5}$ E. $\dfrac{1}{b} < 5 < \dfrac{1}{5}$

12. Solve for y in terms of x: $3x + y > xy$.

ANSWERS:

1. A. False—if a and b are positive and c is negative, then ac < bc

   B. False—if a and b are both negative, then $a^2 < b^2$

   C. True

   D. False—if a and b are positive and c is negative, then $\dfrac{a}{c} < \dfrac{b}{c}$

2. 3x + 2 > 11

   3x > 9

   x > 3

3. -3x + 2 > 11

   -3x > 9

   x < -3

4. $\dfrac{P}{N} > 2$ Therefore, P > 2N (when N>0) or P < 2N (when N<0)

N > 0	N < 0
$\dfrac{P}{N} > 2$	$\dfrac{P}{N} < 2$
P > 2N	P < 2N

5. $+4 < x < +10$

   $-2 < y < +12$

   $\overline{+2 < x + y < +22}$

6. Choice A could not be true for any of the numbers in the range: If a > 1 and you add another positive number to it you will always have a number that is greater than 1.

7. 3x - 8 < x - 18

   2x < -10

   $x < \dfrac{-10}{2}$

   x < -5

8. If, $0 < a - b < 3$

   adding b: $0 + b < a < 3 + b$

   therefore: $b < a < 3 + b$

85

9. If $a > 1$, then $a^2 > a$. Example: $3^2 > 3$.

    If $0 < a < 1$, then $a^2 < a$. Example: $\left(\dfrac{1}{2}\right)^2 < \dfrac{1}{2}$.

10. Since $a < 0$, a is a negative number. Squaring a negative number produces a positive. Therefore $a^2 > a$.

11. The correct answer is B. If $a < 5 < b$, since a and b are positive the reciprocal will reverse the order of the inequality.

12. SOLUTION: $3x + y > xy$ or $xy < 3x + y$

$$y - xy > -3x \qquad\qquad xy - y < 3x$$

$$y(1 - x) > -3x \qquad\qquad y(x - 1) < 3x$$

$$y > \frac{-3x}{1-x} \qquad\qquad y < \frac{3x}{x-1}$$

As you can see, the two answers are equivalent.

Set Theory and Logic

SET THEORY

We have already introduced the set notation in previous lessons.

A = {3, 6, 9} is the set of all numbers between 1 and 10 that are multiples of 3.

B = {6}   is a subset of A. It could represent all numbers between 1 and 10 that are not only multiples of 3, but also multiples of 6.

We use the symbol $B \subset A$ to denote the subset relationship.

The set of all odd integers evenly divisible by 2 could be represented as: { } or ∅. There are no members of this set. It is called the **null** set or **empty** set.

The set itself and the null set are considered subsets of the set.

EXAMPLE: How many subsets are there of set A, if A = {3, 6, 9}?

    (A) 3
    (B) 6
    (C) 8
    (D) 9

SOLUTION: If we just try to list them, we would have:

{3}, {6}, {9}, {3,6}, {3,9}, {6,9}, {3, 6, 9}, and { }, so there are 8 subsets and the correct answer is (C).

In general, the number of subsets is equal to 2^n, where n = the number of members in the set. In our example, n = 3 and thus 2^n = 8 is the number of different subsets. For each of the elements, we have the choice of including it in the subset, or not including it.

SET RELATIONSHIPS

We've already studied one relationship between sets, and that's the subset relationship, where all the members of the subset were also members of the original set. We want to discuss two other relationships, the **union** of two or more sets and the **intersection** of two or more sets.

Union
The **union** of two sets is a set of all elements contained in either of the original sets.

EXAMPLE: If A = {x: x is even and 0 < x < 12} = {2, 4, 6, 8, 10}, and

B = {x: x is a perfect square and 0 < x < 11} = {1, 4, 9},

then the union of these two sets, represented as **A** $\cup$ **B** is all elements that are in either **A or B or both.**

A $\cup$ B = {1, 2, 4, 6, 8, 9, 10}.

Intersection

The **intersection** of two sets is a set of all elements common to both sets.

EXAMPLE: In our previous example, the common elements would be just the number 4. It is the only number between 1 and 10 that is both even and a perfect square. We represent the intersection as: A $\cap$ B = {4}.

EXAMPLE: If A = {positive odd integers < 30} and B = {perfect squares < 30}, then how many elements are there in A $\cup$ B?

(A) 17
(B) 20
(C) 3
(D) 15

SOLUTION: A = {1, 3, 5, 7, 9, 11, 13, 15, 17, 19, 21, 23, 25, 27, 29}

B = {1, 4, 9, 16, 25}

A $\cup$ B = {1, 3, 4, 5, 7, 9, 11, 13, 15, 16, 17, 19, 21, 23, 25, 27, 29}

The correct answer is (A).

By inspection, we know that there are 30 integers between 1 and 30 inclusive, and half of them are odd, or 15. The union would include any perfect squares that were even, or 4 and 16. 15 + 2 = 17.

EXAMPLE: If A = {2, 4, 6, 8, 10}, B = {10, 15, 20, 25}, and C = {3, 6, 9, 12, 15}, then what is A $\cap$ (B $\cup$ C)?

(A) {2, 4, 6, 8, 10, 15}
(B) {10}
(C) {6, 10}
(D) {2, 3, 4, 6, 8, 9, 10, 12, 15, 20, 25}

SOLUTION: {B $\cup$ C} = {3, 6, 9, 10, 12, 15, 20, 25}, any element that is in one or both of the sets.

A $\cap$ (B $\cup$ C) are the elements in common between A and the union of B and C. The common elements are 6 and 10.

The correct answer is (C).

EXAMPLE: Given A = {integers x: x > 10} and B = {integers x: < or = 20}, what is the number of elements in A ∩ B?

(A) 9
(B) 10
(C) 11
(D) 0

SOLUTION: The intersection of these sets is all whole numbers greater than 10 but less than or equal to 20. The number 10 is not an element of the intersection, but the number 20 is. We count the numbers 11 through 20.

The correct answer is 10 or (B).

Complement

The final relationship we want to discuss is **complement**. It deals with the relationship between 2 specific subsets of a given set. If B is a subset of A, the complement of B (written ~B) is a subset containing all the elements of A not included in B.

EXAMPLE: If A = {all integers between 1 and 10 inclusive} and
B = {all even integers between 1 and 10 inclusive}, then the complement of B is {all odd integers between 1 and 10 inclusive}.

VENN DIAGRAMS

We can use circles to represent the relationship between sets. These circle representations are called Venn diagrams. For Example:

A = {4, 8, 12, 16}

B = {6, 8, 12, 20}

C = {4, 12, 20, 28}

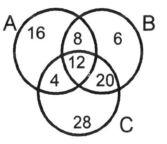

A ∩ B = {8, 12}

A ∩ C = {4, 12}

B ∩ C = {12, 20}

A ∩ B ∩ C = {12}

EXAMPLE: In the Venn Diagram below, the shaded region represents which relationship?

(A) $A \cup B$
(B) $A \cup (B \cup C)$
(C) $A \cup C$
(D) $A \cup (B \cap C)$

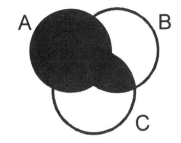

By inspection, we see that we've shaded all of circle A plus the area where circles B and C meet, or intersect. So, we are joining A with the intersection of B and C. The correct answer is (D).

EXAMPLE: Given the Venn Diagram, which of the following statements are true?

I. $A \cup B = \{4, 6, 8\}$

II. $A \cap C = \{5, 6, 7\}$

III. $B \cap C = \{16, 24\}$

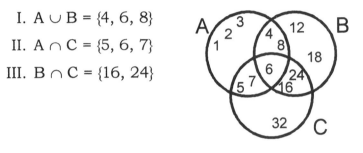

(A) I and II
(B) II only
(C) All
(D) None

SOLUTION: A is obviously false because it doesn't even include all the elements of A.

{4, 6, 8} is the intersection of the two sets. Therefore, we can eliminate a) and c). B is true which eliminates D as a response. The correct answer is B.

The Cartesian product of two sets is the set of all ordered pairs that pairs each element from the first set with each element from the second set. The Cartesian product is shown by A X B.

EXAMPLE: If A = {1, 2, 3} and B = {a, b},

then A X B = {(1, a), (1, b), (2, a), (2, b), (3, a), (3, b)}.

Notice that in every pair the first element is from the first set, the second from the second set. The important point is the order.

EXAMPLE: Thus, for the same sets,
B X A = {(a, 1), (b, 1), (a, 2), (b, 2), (a, 3), (b, 3)}.

These are different ordered pairs than in A X B. Can you see that (A X B) ∩ (B X A) =∅? Note that if A has 3 elements, and B has 2 elements, then A X B has 3 × 2 = 6 ordered pairs. This is a simple but important counting principal.

Closure

Closure is a property of sets with respect to an operation. For example, the set of all integers is **closed** with respect to addition. This means that the result of adding any two members of the set of integers is also a member of the set of integers. We know this is true because we know that adding two integers results in an integer. The set of all integers is also closed for multiplication. The terms "closure" and "closed" allow us to say a lot quickly. A more generic statement is, "If **a** and **b** are members of the set **X** then the set **X** is closed with respect to operation **f(a,b)** if **f(a,b)** is in the set **X** for all possible **a** and **b**."

Closure, a property of sets, regards the effect of a certain operation on the elements of a set. Closure asks for a <u>given operation</u> (such as addition, subtraction, multiplication, or division) <u>performed on a given set</u> if all results are within the given set. If any result is outside the set, the set is not closed. If all results are within the set, the set is closed.

EXAMPLE: The set {1, 0} is closed under multiplication since all the answers are in the set: 1 • 1 = 1, 1 • 0 = 0, 0 • 0 = 0.

NOTE: the operation (multiplication here) can be performed on an element with itself.

For the same set, {1, 0}, the set is not closed under addition, since 1 + 1 = 2 and 2 is not an element of the set.

EXERCISE: Is the set of even numbers closed under addition?

The answer is yes, since adding even numbers gives an even number which, therefore, must be in the given set. How about the set of odd numbers under addition?

The concept of sets is an extremely useful tool in many areas of mathematics. Many of its uses, however, are beyond the scope of this book and the CLEP examination. Be sure to learn the symbols, and you will find that set concepts are not difficult.

A favorite type of set problem which uses set symbolism might be like this:

EXAMPLE: A = {x: x ≥ 1}, B = {x: x ≤ 1}. Find A ∩ B.

This is read: A is the set of numbers x, such that x is greater than or equal to 1. B is the set of numbers x, such that x is less than or equal to 1. Find the intersection of sets A and B.

SOLUTION: Since A is the set of numbers from 1 up and B is the set of numbers from 1 down, the only common element is 1. Therefore, $A \cap B = \{1\}$.

NOTE: Sometimes the symbol | is used in place of the colon to mean "such that." $A = \{x \mid x \geq 0\}$ says A is the set of elements x such that each element is greater than or equal to zero.

By trying the set exercises in this study guide, you will familiarize yourself with set symbols and find that set problems are ones which you can easily solve and are ones with which you can readily earn valuable points.

PRACTICE PROBLEMS — SET THEORY

Questions 1 - 7 refer to the following:

$A = \{1, 2, 3, 4, 5\}$

$B = \{2, 3, 5, 8, 9\}$

$C = \{2, 4, 6, 7, 8\}$

1. What is $A \cup B$?

   (A) {2,3}
   (B) {2,3,5}
   (C) {1,4,8,9}
   (D) {1,2,3,4,5,8,9}

2. What is $A \cap B$?

   (A) {1,2,3,4,5,8,9}
   (B) {∅}
   (C) {2,4}
   (D) {2,3,5}

3. What is $A \cup (B \cap C)$?

   (A) {2,3,5}
   (B) {1,2,3,4,5,8}
   (C) {2,3,4,5}
   (D) {2,3,4,5,6,7,8,9}

4. How many subsets of C are there?

   (A) 5
   (B) 7
   (C) 32
   (D) 16

5. Which of the following is **not** an element of B X C?

   (A) (3,4)
   (B) (2,6)
   (C) (2,8)
   (D) (3,9)

6. How many pairs are in the set A X C?

   (A) 5
   (B) 10
   (C) 25
   (D) 7

7. What is (A ∩ B) ∩ C?

   (A) {2}
   (B) {2,3,5}
   (C) {∅}
   (D) {2,3,4,5,6,8}

8. Which of the following sets is closed under addition?

   (A) {0,1}
   (B) {-1,0,1}
   (C) {even numbers}
   (D) {odd numbers}

9. If A = {x: -1 < x ≤ 3} and B = {X: x ≥ 3}, then A ∪ B =?

   (A) x < -1
   (B) x > -1
   (C) x ≥ 3
   (D) x ≥ -1

10. In the Venn diagram below, the shaded region represents which of the following?

(A) $A \cup B \cup C$
(B) $A \cap B \cap C$
(C) $(B \cup C) \cap A$
(D) $(A \cap C) \cup C$

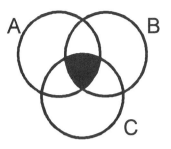

11. Based on the Venn diagram below, which of the following statements is true?

(A) $A \cup B = \{2,4\}$
(B) $A = \{1,2,3,4,6,8\}$
(C) $A \cap B = \{2,4\}$
(D) $B = \{1,2,3,4,6,8\}$

ANSWERS

1. D $A \cup B$ contains all of the elements included in either A or B.

2. D $A \cap B$ contains only those elements common to both sets.

3. B   The intersection of B and C is {2,8}. The union of {2,8} with A = {1,2,3,4,5} is {1,2,3,4,5,8}.

4. C   There are 5 elements so there are 2^5 or 32 subsets including the null set and the set itself.

5. D   The first element must be from set B and the second element must be from set C. In choice D, both elements are from B.

6. C   Each of the 5 elements in set A can be paired with each of the 5 elements in set B. A X B has 5 x 5 or 25 ordered pairs.

7. A $A \cap B$ is {2,3,4}; {2,3,4} $\cap$ {2,4,6,7,8} is {2}.

8. C   Since adding even numbers produced an even sum, all answers are within the given set. The set of even numbers is, therefore, closed under addition.

9. B   A is the set of numbers greater than -1 and less than or equal to 3. B is the set of numbers greater than or equal to 3. $A \cup B$ is the set of numbers greater than -1.

10. B   The shaded part represents the elements common to A and B and C or $A \cap B \cap C$. Using the elements defined in problems 1 - 7, the intersection of the 3 sets would be {2}.

11. C   The intersection of A and B includes those elements common to both sets, or in the Venn diagram those elements included in the area where the 2 circles intersect. $A \cap B = \{2,4\}$.

LOGIC

An area of mathematics which utilizes many of the principles in set theory is logic. As in sets, the key to these problems is learning the symbols which are used. There are no real computations. Rather, logic tests your reasoning powers in problems in which English sentences have been replaced with **symbols**.

The basic unit of logic is the statement—a simple sentence, such as, "I am tall." In logic we are concerned with the compounding of simple statements. To do this we replace statements with letters.

EXAMPLE: Let p represent the statement, "I am tall," and q represent the statement, "I am a good tennis player." The compound sentence, "I am tall, and I am a good tennis player," could be written as p and q. To simplify even further, we replace "and" by the symbol ∧. Now we can write p ∧ q. This is known as the **conjunction**.

To write, "I am tall or I am a good tennis player," we write p ∨ q. This is known as the **disjunction** of p and q. To write, "I am not tall," we use ~p. This is known as the **negation** of p. The negation of p ∧ q would be ~(p ∧ q). This would translate as, "I am not both tall and a good tennis player." How would ~p ∨ q translate? "I am not tall, or I am a good tennis player."

A **truth table** shows the truth or falsity of a statement using the symbols of logic.

EXAMPLE:

p	~p
T	F
F	T

This table says:

If p is true, ~p is false.

If p is false, ~p is true.

The truth table for a compound sentence depends upon the truth of its components. The truth table for p ∧ q would be:

p	q	p ∧ q
T	T	T
T	F	F
F	T	F
F	F	F

If both p and q are true, p ∧ q is true.

If p is true but q is false, p ∧ q is false.

If p is false and q is true, p ∧ q is false.

If both p and q are false, p ∧ q is false.

The truth table for p ∨ q would be:

p	q	p ∨ q
T	T	T
T	F	T
F	T	T
F	F	F

If both p and q are true, p ∨ q is true.

If either p or q is true, p ∨ q is true.

If either p or q is true, p ∨ q is true.

If both p and q are false, p ∨ q is false.

Now let's take a look at two statements:

A: You have a press pass.

B: You can join the team on the field.

Combining statements A and B, we make the following sentence:

If you have a press pass, then you can join the team on the field.

This is called a conditional statement. It has a **hypothesis or antecedent**, "If you have a press pass" and a **conclusion or consequent**, "then you can join the team on the field". The conclusion that you join the team on the field is conditional. It won't always happen. Not everyone gets to go onto the field. And, I don't know if you have a pass. But let's just "hypothesize" that you do have one. Then logically I would conclude that you can join the team on the field.

Symbolically, we represent this conditional relationship as A → B. This can be read as "A implies B" or "If A, then B".

In conditional statements we also talk about **necessary** and **sufficient** conditions. In our example, having a press pass is **sufficient** for joining the team on the field. However, it's not **necessary**. You could be the coach or the ref, etc.

Suppose we reversed these two statements. The statement "If B, then A" or B → A is called the **converse** of "If A, then B". In our example, that would be "If you join the team on the field, then you have a press pass". As you can see, the converse is not necessarily true. Remember, it is sufficient but not necessary.

The statement "If not A, then not B" or ~A → ~B is called the **inverse** of "If A, then B". In our example, that would be "If you don't have a press pass, you can't join the team on the field". As with the converse, we know that this is not necessarily true.

Finally, the statement "If not B, then not A" or ~B → ~A is called the **contrapositive** of "If A, then B". In our example, that would be "If you can't join the team on the field, then you don't have a press pass". If the original statement is true, and in our example, it is, then the contrapositive is also true. If you are not on the field with the team, that tells us a lot. It tells us that you are not on the team, that you are not the coach, that you are not a ref, but it also tells us that you don't have a press pass. Because, if you did have a pass, you would be there on the sideline, even if you were not the coach or a ref.

In summary:

The original conditional statement is written A → B.

The **converse** is written B → A.

The **inverse** is written ~A → not ~B.

The **contrapositive** is written ~B → ~A.

The original statement and its contrapositive are logically equivalent. If the original statement is true, then the contrapositive will also be true. If the original statement is false, then its contrapositive will also be false.

EXAMPLE: The conditional statement "If you are not dressed, you can't go outside" is logically equivalent to which of the following?

(A) If you can't go outside, then you aren't dressed.
(B) If you are dressed, then you can go outside.
(C) If you aren't dressed, then you can go outside.
(D)   If you can go outside, then you are dressed.

SOLUTION: If we let A equal "You are dressed", and B equal "You can go outside", then the original statement can be represented as: ~A → ~B. This is the contrapositive of and is logically equivalent to the statement B → A. This is the statement "If you can go outside, then you are dressed". The correct answer is (D).

EXAMPLE: A: You are a woman.

B: You can't become President of the USA.

SOLUTION: The conditional statement A → B would be "If you are a woman, then you can't become President". This is false. If the hypothesis is true and the conclusion is false, then the conditional statement is false.

This is the only time that a conditional statement is logically false. If the hypothesis and conclusion are both true, then the conditional statement is true. Also, if the hypothesis is false, regardless of the conclusion, the conditional statement will be logically true.

EXAMPLE: A: The moon is made of green cheese.

B: The pictures of men walking on the moon are fakes.

SOLUTION: A is a false statement, but the conditional statement A → B, or "If the moon is made of green cheese, then the pictures of men walking on the moon are fakes" is logically true. I'm not saying that it's true that the pictures are fake. I'm saying that the conclusion logically follows and that the conditional statement is true. Think about that for a second. As remote as the possibility might seem, if it turns out that the moon really is made of green cheese, then we really have been duped.

So, to reiterate, the only time a conditional statement is logically false, is when the hypothesis is true but the conclusion is false.

p	q	p → q	
T	T	T	If both antecedent and consequent are true, the conditional is true.
T	F	F	If the antecedent is true but the consequent is false, the conditional is false.
F	T	T	If the antecedent is false but the consequent is true, the conditional is true.
F	F	T	If both antecedent and consequent are false, the conditional is true.

Now, getting back to our original example:

A: You are a woman.

B: You can't become President of the USA.

The **contrapositive** is "If not B, then not A," or ~B → ~A. In our example, that would be "If you can become President, then you are not a woman". Again, the hypothesis is true but the conclusion is false, so the conditional statement is false.

The converse and the inverse have a similar logical relationship.

Now let's look at the statement:

"You can go to the game on Sunday if and only if you help rake the leaves on Saturday."

Let A represent the statement "You can go to the game on Sunday"; and let B represent the statement "You help rake the leaves on Saturday". If we know that you can go to the game on Sunday, then we conclude that you must have raked the leaves on Saturday.

If A → B. Also, if we know that you raked leaves on Saturday, then we can conclude that you can go to the game on Sunday. If B → A. Raking the leaves is not only sufficient for going to the game, it is also necessary. No other good deed, no amount of money will make it possible for you to attend the game. So, if I hypothesize that you do, in fact, rake the leaves, then I can logically conclude that you can go to the game. Also, if I hypothesize that you can go to the game, then I can logically conclude that you must have raked the leaves.

An "if" statement implies a sufficient condition. An "only if" statement implies a necessary condition.

EXAMPLE: Given the statement, "If Christmas is tomorrow, then it must be December", which of the following statements is the inverse:

(A) If tomorrow isn't Christmas, then it isn't December.
(B) If it is December, then tomorrow is Christmas.
(C) If it isn't December, then tomorrow isn't Christmas.
(D) If Christmas is tomorrow, then it isn't December.

SOLUTION: A: Christmas is tomorrow.

B: It is December.

The inverse of If A, then B, is If ~A, then ~B. Or in this case, If tomorrow isn't Christmas, then it isn't December. The correct answer is (A).

(B) is If B, then A, which is the converse.

(C) is If ~B, then ~A, which is the contrapositive.

PRACTICE PROBLEMS — LOGIC

1. If a represents "It is raining" and b represents "The streets are wet", which of the following represents the statement, "It is raining and the streets are wet"?

(A) a ∨ b
(B) a ∧ b
(C) a → b
(D) a → ~b

2. Which of the following is the converse of a → b?

   (A) a → ~b
   (B) b → a
   (C) ~a → b
   (D) a ∧ b

3. Which of the following is the negation of p ∨ q?

   (A) p → q
   (B) q → p
   (C) ~p ∧ q
   (D) ~(p ∨ q)

4. If p ∨ q is false, then which of the following must be false?

   (A) p
   (B) q
   (C) both p and q
   (D) neither p or q

5. Which of the following is the inverse of a → b?

   (A) ~a → ~b
   (B) a → ~b
   (C) b → a
   (D) ~a → b

6. Which of the following is the contrapositive of a → b?

   (A) b → a
   (B) ~a → b
   (C) ~a → ~b
   (D) ~b → ~a

7. If p represents "I am a good student" and q represents "I will pass the test", which of the following represents "If I am not a good student, I will not pass the test"?

   (A) p → q
   (B) ~p → ~q
   (C) ~q → ~p
   (D) q → p

8. If the statement "All lawyers know Latin" is true, which of the following statements is implied from the first statement?

(A) All non-lawyers do not know Latin.
(B) If a person knows Latin, he is a lawyer.
(C) If a person does not know Latin, he is a lawyer.
(D) If a person does not know Latin, he is not a lawyer.

9. If p is true and q is false, then p ∨ q is:

(A) false
(B) true
(C) equals p ∧ q
(D) conditionally false

10. Given the conditional statement, p → q, then q → p is

(A) the converse
(B) the inverse
(C) the contrapositive
(D) always true

ANSWERS

1. B   The symbol ∧ is called the conjunction and replaces the word "and".

2. B   The converse switches the antecedent and the consequent.

3. D   The symbol ~ is used for negation.

4. C   Only when p and q are both false will the disjunction be false. In other words, if either one statement or the other is true, we consider the statement true.

5. A   The inverse is the negation (~) of the conditional a → b.

6. D   The contrapositive is the negation of the converse.

7. B   The statement represents the inverse.

8. D   Only statement D is true if the original statement is true.

9. B   If either p or q is true, then the disjunction is considered true.

10. A   q → p is the converse of the conditional p → q. The converse is not always true.

Coordinate Geometry

Coordinate or Analytical Geometry deals with location and relationship of points located on a two-dimensional surface. This surface is divided into 4 parts by two intersecting perpendicular lines called axes. The horizontal line is the x-axis and the vertical line is the y-axis. They are like horizontal and vertical number lines. Every point in this space is defined by an ordered pair of numbers, where the first number is always the distance traveled along the x-axis and the second number is the distance traveled along the y-axis. The point where the x-axis and y-axis intersect is (0,0) and is called the **origin.**

Locate the points (5,2) and (-1,1).

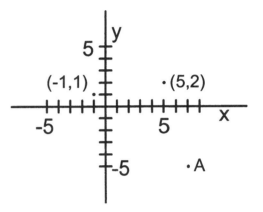

The point (5,2) is located by moving to the right 5 spaces from the origin and then up 2 spaces.

The point (-1,1) is located by moving to the left 1 space from the origin and then up 1 space.

EXAMPLE: What are the coordinates of point A?

    (A) (7,5)   (B) (7,-5)

    (C) (-7, 5)   (D) (-7,-5)

SOLUTION: To locate point A, you must move to the right 7 units from the origin and then down 5 units. That's a +7 and a -5. The point is (7, -5). The correct answer is (B).

GRAPHING A STRAIGHT LINE

We can learn important information about a function by graphing it. The simplest way to graph a function is to find ordered pairs by substituting different values for the variables and finding the value of the function. We can then make a table and plot the points of the function.

To graph the algebraic equation y = 2x, we can set up a table of values by substituting values for x and solving for values of y:

x	y
0	0
1	2
2	4

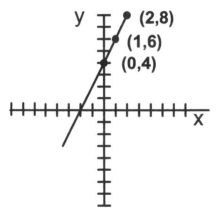

We can plot these points and connect them with a straight line as shown in the graph to the right of the table.

EXAMPLE: If f(x) = 2x + 4, then f(2) = 8, f(1) = 6, f(0) = 4, etc.

By plotting the points (2, 8), (1, 6), (0, 4), we can get a picture of the graph. As you can see below, this function is a straight line:

What we notice when we graph a straight line is that there is a constant relationship between the x and y coordinates, and in general, straight lines are represented by the equation y = mx + b where m is the slope of the line, and b is the y-intercept. y is a **linear** function of x.

The y-intercept is the point where the line crosses the y-axis, or where x = 0.

In our equation, if we substitute 0 for x, we see that y = b.

The slope defines the slant of the line. In the equation y = 2x, we see that each time we move 1 in an x direction, we will move 2 in a y direction. m, written as a fraction, represents the change in y over the change in x. The larger the number, the steeper the slant. Positive slopes rise from left to right. Negative slopes fall from left to right.

EXAMPLE: What is the equation of a line that passes through (0,7) and (6,4)?

(A) $y = \dfrac{1}{2} x$

(B) $y = 2x + 7$

(C) $y = 2x$

(D) $y = -\dfrac{1}{2} x + 7$

SOLUTION: The slope is change in y over change in x, so slope is $\dfrac{4-7}{6-0} = -\dfrac{1}{2}$. So, the equation is in the form $y = -\dfrac{1}{2}x + b$. Since either point must make this a true statement, we can substitute either in to solve for b.

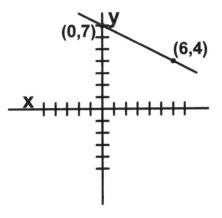

$$4 = -\dfrac{1}{2}(6) + b$$

$$4 = -3 + b$$

$$7 = b$$

The equation is $y = -\dfrac{1}{2}x + 7$.

The correct answer is (D).

By inspection, the problem gave us the y-intercept, it's the value when x is 0, so we know that the value of b is 7. This eliminates (A) and (C). We also know that y decreases while x increase, so that the slope is negative. As an alternative, we see that the change in x is greater than the change in y, so we know that the slope is less than 1. Either way we eliminate (B). It has a positive slope greater than 1.

EXAMPLE: If f(x) is a linear function such that f(-4) = 5 and f(2) = 8, then f(0) =

(A) 4

(B) 7

(C) 5

(D) 9

106

SOLUTION: The slope of the line (m) is equal to the change in y over the change in x. This is equal to $\dfrac{5-8}{-4-2} = \dfrac{-3}{-6} = \dfrac{1}{2}$. So, the equation of our line is equal to $f(x) = \dfrac{1}{2}x + b$. We know that $f(2) = 8$, so b must equal 7. Our equation is $f(x) = \dfrac{1}{2}x + 7$. $f(0) = 7$. The correct answer is (B).

GRAPHING QUADRATIC FUNCTIONS

The graph of a function in which the variable has an exponent of 2 usually yields a curve, such as a circle, a parabola, an ellipse, or a hyperbola. By plotting enough pairs of points, you can get a rough "sketch" of the curve.

EXAMPLE: To graph $y = f(x) = x^2 + 2$:

x =	0	1	-1	2	-2
y =	2	3	3	6	6

As you can see, the same number for x positive or negative gives equal values of y. Plotting the points, we would get a rough sketch as shown below:

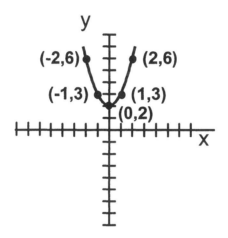

There are some important parts on any graph. The point where the line or curve crosses the x-axis is called the **x-intercept**. The point where a curve or line crosses the y-axis is the **y-intercept**.

EXAMPLE: You can see in the function $f(x) = x^2 + 2$, the curve is always above the x-axis; therefore, it has no real number x-intercepts. The y-intercept is the point when $x = 0$. In $f(x) = x^2 + 2$, this point is (0, 2).

The point where two functions cross is an ordered pair (coordinates), which satisfy both functions. That is, substituting the first number for x and the second for y will make true statements for both functions.

107

EXAMPLE: If the two functions y = f(x) = x² + 1 and the function y = f(x) = x + 3 are both graphed, they will cross at the points (2, 5) and (-1, 2). Here is the graph of y = f(x) = x² + 1 and y = f(x) = x + 3:

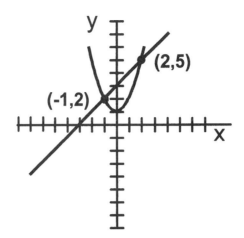

The both pairs satisfy both functions as you can see:

for:	y = x² + 1			y = x+ 3	
(2, 5)	5 = 2² + 1	True	(2, 5)	5 = 2 + 3	True
(-1, 2)	2 = (-1)² + 1	True	(-1, 2)	2 = -1 + 3	True

The two most important concepts when graphing functions are: (1) if a point lies on a line or curve of a graph, its coordinates must satisfy the equation, and (2) if the coordinates of a point do satisfy the equation, then the point must lie on the graph.

If two functions cross at two points that must satisfy both functions, we can solve algebraically for these points by setting the functions equal to each other.

EXAMPLE: Using the examples from above, if y = f(x) = x² + 1 and
y = f(x) = x + 3

SOLUTION: Then, x² + 1 = x + 3 and x² - x - 2 = 0.

Factoring: (x - 2)(x + 1) = 0

x – 2 = 0   x + 1 = 0

x = 2   x = -1

Then y = 2² + 1 = 5 and y = 2 + 3 = 5 and y = (–1)² + 1 = 2 and y = -1 + 3 = 2.

These points (2, 5) and (-1, 2) are the points of intersection of the two functions.

PRACTICE PROBLEMS — COORDINATE GEOMETRY

1. Which of the following numbered graphs could represent the graph of
    $y = 2x+4$?

    (A) 1
    (B) 2
    (C) 3
    (D) 4

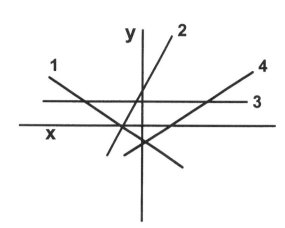

2. Which of the following functions would graph as a curve?

    I. $y = x^2 + 2$

    II. $y = 2x$

    III. $y = 3(x - 2)$

    (A) I only
    (B) I and II
    (C) I and III
    (D) III only

3. Which of the following functions fits the ordered pairs shown?

    $f(x, y) = \{(0, 1), (2, 5), (4, 17)\}$

    (A) $f(x) = 2x$
    (B) $y = x + 3$
    (C) $f(x) = x^2 + 1$
    (D) $f(x) = x^2$

ANSWERS

1. B   For $y = 2x + 4$, the slope is 2 and the y-intercept is 4. Only line 2 has a positive slope and a positive y-intercept. Equations 1 and 4 have negative y-intercepts and equation 3 has a slope of zero.

2. A   Function I is a curve. Functions II and III are linear (straight-line) equations.

3. C   By substituting the first number for x in each function, you find only
$x^2 + 1$ gives the second element of the ordered pair. For example,
$17 = 4^2 + 1$

Probability

PERMUTATIONS

The **permutations** of a set of objects are the number of different subsets you can form from a set, <u>where order is important</u>.

EXAMPLE: How many different ways can I arrange a set of 6 different colored flags?

SOLUTION: I have six different ways that I can fill the first position. Once I've filled the first position, there are 5 flags left, so I have 5 choices for the second position, 4 for the third, and so on.

6* 5 * 4 * 3 * 2 * 1

Each way I can fill the first position is paired with each way I can fill the second position, etc. This pairing implies multiplication, and, in fact, the total number of permutations is

6 * 5 * 4 * 3 * 2 * 1 or 720 permutations.

This product can also be written as **6!** (called 6 factorial).

EXAMPLE: From the same set of 6 different colored flags, how many different arrangements of three flags can I make?

SOLUTION: 6 * 5 * 4 = 120

This could also be written as $\dfrac{6 \cdot 5 \cdot 4 \cdot 3 \cdot 2 \cdot 1}{3 \cdot 2 \cdot 1} = \dfrac{6!}{3!}$

In general, the formula for the permutations of n things taken r at a time, or $_nP_r$, is $\dfrac{n!}{(n-r)!}$, where n equals the total number of objects and r equals the number chosen.

EXAMPLE: Marisa bought 5 new blouses, 3 new skirts, and 2 new pairs of shoes for her new job. If she works Monday through Friday, how many weeks can she go without wearing the same outfit twice?

(A) 1
(B) 3
(C) 6
(D) 2

SOLUTION: First, we have to find how many different outfits can she make. She has 5 choices for the blouse, which can be paired with any one of 3 choices for the skirt, which, in turn, can be paired with either of the 2 choices for the shoes.

5 * 3 * 2 = 30

So, she can make 30 different outfits. But how many weeks can she go without repeating an outfit? That would be $\frac{30}{5}$ or 6 weeks.

The correct answer is (C).

COMBINATIONS

Combinations are the number of different subsets you can form from a set of objects, <u>where order is not important</u>.

EXAMPLE: The Senior Class Council is made up of one representative from each of the 12 Senior Homerooms. They are going to pick 3 members to serve as a subcommittee to plan the Senior Prom. How many different subcommittees are possible?

(A) $\dfrac{12!}{9!3!}$

(B) $\dfrac{12!}{9!}$

(C) $\dfrac{12!}{3!}$

(D) $\dfrac{12!}{9!(9-3)!}$

SOLUTION: If order were important in the selection we would have 12 * 11 * 10. But that would be counting every three-member panel 3! times. The committee of Fred, Sally, and George and the committee of Sally, Fred, and George would be considered two different committees if order were important. However, we know that this is the same committee, regardless of the order in which we picked the 3 names. So, we must divide 12 * 11 * 10 by 3! to eliminate the repetition.

The number of subcommittees $= \dfrac{12 \cdot 11 \cdot 10}{3!} = \dfrac{12 \cdot 11 \cdot 10 \cdot 9!}{3!9!} = \dfrac{12!}{3!9!}.$

The correct answer is (A). When multiplied it out, it is 220.

In general, the formula for the Combination of n things taken r at a time, or $_nC_r$, is:

$$_nC_r = \dfrac{n!}{r!(n-r)!}$$

EXAMPLE: How many different hands of 5 cards can you deal from a 52-card deck?

SOLUTION: The order that you receive the cards is unimportant. The only thing you care about what 5 cards you have at the end of the deal.

We already know that if order were important, we have $_{52}P_5 = \dfrac{52!}{47!}$.

What's the impact of considering order? I'm counting each hand of 5 cards 5! times. If I divide the number of permutations by 5!, I'll have the number of combinations.

$$_{52}C_5 = \frac{52!}{47! * 5!} = \frac{52 * 51 * 50 * 49 * 48 * 47!}{47! * 5!} = \frac{52 * 51 * 50 * 49 * 48}{5 * 4 * 3 * 2 * 1}$$
$$= 13 * 17 * 10 * 49 * 24 = 2,598,960$$

PROBABILITY

The probability that an event will occur is the ratio of the number of favorable outcomes to total number of outcomes.

EXAMPLE: If you roll a die, what is the probability that you will roll an even number?

SOLUTION: The set of favorable outcomes = {2, 4, 6}

The set of total outcomes = {1, 2, 3, 4, 5, 6}

$$P(even) = \frac{3}{6} = \frac{1}{2}$$

EXAMPLE: What is the probability that you draw an Ace from a 52-card deck?

SOLUTION: There are four favorable outcomes = the Ace of spades, hearts, diamonds, & clubs. There are 52 total outcomes.

$$P(Ace) = \frac{4}{52} = \frac{1}{13}$$

The probability of not getting an ace is $\frac{48}{52}$ or $\frac{12}{13}$. The probability that an event doesn't occur is equal to 1 - P(Event).

EXAMPLE: What is the probability that a single draw from a standard 52-card deck will yield a number divisible by 3?

(A) $\frac{3}{52}$

(B) $\frac{1}{13}$

(C) $\frac{3}{13}$

(D) $\frac{2}{13}$

SOLUTION: There are 3 numbers divisible by 3 (3, 6, and 9) and there are 4 of each in the deck, so there are 12 favorable outcomes. There are 52 total outcomes. The probability is $\frac{12}{52}$ or $\frac{3}{13}$. The correct answer is (C).

115

EXAMPLE: Bob has 7 shirts, each a different color. He randomly selects 3 shirts to pack for a business trip. What is the probability that he packs the blue, yellow, and white shirts?

(A) 210

(B) $\dfrac{1}{840}$

(C) $\dfrac{1}{35}$

(D) $\dfrac{1}{210}$

SOLUTION: $_7C_3 = \dfrac{7!}{3!4!} = \dfrac{7*6*5}{3*2*1} = 7*5 = 35$

There are 35 possible combinations of shirts, only 1 of which is blue, yellow, and white. There is only one favorable outcome out of 35 total outcomes. The probability is $\dfrac{1}{35}$.

(C) is the correct answer.

PROBABILITY OF A COMBINATION OF EVENTS

We've discussed the probability of a single event occurring. What about the probability that one of two independent events occur? Either event A or event B. The word "or" in probability translates to addition.

EXAMPLE: What is the probability that you draw an Ace or a King from a 52-card deck?

SOLUTION: This is equal to the probability that you draw an ace plus the probability that you draw a king.

P(Ace or King) = P(Ace) + P(King) = $\dfrac{1}{13} + \dfrac{1}{13} = \dfrac{2}{13}$.

The key here is that a card couldn't be both an Ace and a King.

In a situation such as the probability that the card is an Ace or a heart, you would be better off counting the favorable outcomes. There are 4 aces, one of which one is a heart, and there are 12 other hearts. So, there are a total of 16 favorable outcomes out of a total of 52 outcomes. The probability that a card is either an Ace or a Heart, P(Ace or Heart), is equal to $\dfrac{16}{52}$, or $\dfrac{4}{13}$

The word "and" in probability translates to multiplication.

EXAMPLE: If you deal 2 cards from a 52 card deck, what is the probability that both are Aces?

SOLUTION: This is equal to the probability that you draw an ace on the first card times the probability that you draw an ace on the second card (given that you already drew an ace on the first card).

P(2 Aces) = P(Ace on first card) * P(Ace on second card) =

$$\frac{4}{52} * \frac{3}{51} = \frac{12}{2652}$$

The key here is that the first card is not replaced in the deck, so there are only 3 aces remaining for the second card and only 51 cards remaining in the deck.

EXAMPLE: The probability of dealing an Ace or a Heart from a standard 52 card deck is:

(A) $\frac{17}{52}$

(B) $\frac{4}{13}$

(C) $\frac{1}{4}$

(D) $\frac{1}{52}$

SOLUTION: There are 13 hearts, one of which is the ace of hearts. There are 3 other aces that are not hearts. So there are a total of 16 favorable outcomes out of a total of 52. So, the probability of dealing an ace or a heart is $\frac{16}{52} = \frac{4}{13}$. The correct answer is (B).

$\frac{17}{52}$ counts the ace of hearts twice; $\frac{1}{4}$ or $\frac{13}{52}$ only counts the favorable outcomes that are hearts; $\frac{1}{52}$ is the probability that the card is an ace **AND** a heart.

EXAMPLE: A fair die is rolled and a card is drawn from a standard deck of 52 cards. What is the probability that the top face of the die will show an even number and the card will be a club?

(A) $\frac{1}{8}$

(B) $\frac{3}{4}$

(C) $\frac{1}{2}$

(D) $\frac{1}{4}$

SOLUTION: This is the probability of a combination of events, so we will multiply the individual probabilities together. $P(\text{Even}) = \frac{3}{6} = \frac{1}{2}$. $P(\text{Club}) = \frac{13}{52} = \frac{1}{4}$. $P(\text{Even and Club}) = \frac{1}{2} * \frac{1}{4} = \frac{1}{8}$. The correct answer is (A).

CONDITIONAL PROBABILITY

Conditional Probability is the probability of one outcome given or based on successfully completing or satisfying another condition or event.

EXAMPLE: In a recent exit poll, 6 out of 10 women said that education was a more important issue than taxes, but only 3 out of 10 men agreed. What is the probability that the person surveyed thought education was the more important issue, given that the person was a woman?

SOLUTION:

	E	T	
M	3	7	10
W	6	4	10
	9	11	20

Given that the person is a woman, there are 10 total outcomes, of which 6 would be considered successful. The probability is $\frac{6}{10}$ or $\frac{3}{5}$.

EXAMPLE: Based on the information above, given that a person believes taxes is the more important issue, what is the probability that the person is a man?

(A) $\frac{7}{11}$

(B) $\frac{7}{10}$

(C) $\frac{1}{3}$

(D) $\frac{3}{10}$

SOLUTION: There are 11 favorable outcomes (people believing that taxes is the more important issue. Of these, 7 are men. The probability is $\frac{7}{11}$.

This can be written as P(M | T) = $\frac{7}{11}$. The correct answer is (A).

$\frac{7}{10}$ is the probability that the person believes taxes is the more important issue, given the person is a man.

$\frac{1}{3}$ is the probability that the person is a man, given that the person believes education is the more important issue.

$\frac{3}{10}$ is the probability that the person believes education is the more important issue, given the person is a man.

In general, we write conditional probability, the probability of X given Y as P(X | Y) = n(X∩Y)/n(Y). In our example, the number of people who are men and believe taxes is the primary issue is 7. The number of people who think taxes is the primary issue is 11.

PRACTICE PROBLEMS

1. What is the probability of drawing a picture card from a 52-card deck?

   (A) $\frac{1}{13}$

   (B) $\frac{4}{13}$

   (C) $\frac{1}{52}$

   (D) $\frac{3}{13}$

2. What is the probability that you can flip a coin three times and get heads each time?

   (A) $\frac{1}{8}$

   (B) $\frac{1}{2}$

   (C) $\frac{1}{4}$

   (D) $\frac{3}{8}$

3. In a game of 21 (blackjack) you have a 10 and a 3. What is the probability that your next card will give you a total of 19, 20, or 21 (assume the 10 and 3 are the only cards drawn from the deck so far)?

   (A) $\frac{3}{13}$

   (B) $\frac{6}{25}$

   (C) $\frac{3}{52}$

   (D) $\frac{3}{50}$

4. What is the probability that it will not rain on a given day if the probability that it will rain is 1/3?

   (A) $\frac{1}{3}$

   (B) $\frac{1}{6}$

   (C) $\frac{2}{3}$

   (D) $\frac{1}{2}$

5. If you draw one of 24 slips of paper numbered consecutively from 1 - 24, what is the probability of drawing a number exactly divisible by 3?

(A) $\dfrac{2}{3}$

(B) $\dfrac{1}{24}$

(C) $\dfrac{5}{8}$

(D) $\dfrac{1}{3}$

6. What is the probability that a random draw of a card from a deck is a 9 or 8?

(A) $\dfrac{8}{13}$

(B) $\dfrac{2}{13}$

(C) $\dfrac{1}{13}$

(D) $\dfrac{4}{13}$

7. If you are to choose a committee of 4 people from a group of 8, how many different committees are possible?

(A) 100
(B) 32
(C) 64
(D) 70

8. How many different telephone area codes can be formed using the digits 0 - 9, if 0 isn't used for the first digit?

   (A) 500
   (B) 1000
   (C) 900
   (D) 729

9. How many triangles can be formed from a set of 8 different points, no three of which lie in a straight line?

   (A) 336
   (B) 60
   (C) 70
   (D) 56

10. How many even three-digit numbers are there?

   (A) 500
   (B) 400
   (C) 4450
   (D) 100

ANSWERS

1. D   There are 12 picture cards in a deck. Therefore, there are $\dfrac{12}{52}$ or $\dfrac{3}{13}$ chances of getting a picture card.

2. A   The probability of getting heads 3 times would be $\dfrac{1}{2} \times \dfrac{1}{2} \times \dfrac{1}{2}$ or $\dfrac{1}{8}$.

3. B   If the next card drawn were a 6, 7, or 8, you would have a total of 19, 20, or 21. Since there are 4 of each, there are a total of 12 favorable cards. Since there are 50 cards left, the probability would be $\dfrac{12}{50}$ or $\dfrac{6}{25}$.

4. C.   The probability that an event will happen, plus the probability it will not, must give a sum of 1. Therefore, the probability it will not happen is $1 - \dfrac{1}{3}$ or $\dfrac{2}{3}$.

5. D   There are 8 numbers from 1-24 which are divisible by 3. The probability is, therefore, $\dfrac{8}{24}$ or $\dfrac{1}{3}$.

6. B The probability of getting a 9 or 8 is $\dfrac{8}{52}$, or $\dfrac{2}{13}$.

7. D The number of committees would be $_8C_4 = \dfrac{8 \times 7 \times 6 \times 5}{4 \times 3 \times 2 \times 1}$ or 70.

8. C For the first digit, you have a choice of 9 digits (1-9). For the second and third digits you have a choice of 10 digits each. Therefore, there are $9 \times 10 \times 10$ or 900.

9. D Each triangle requires 3 points. From 8 points there are $_8C_3$ possible triangles. $_8C_3 = \dfrac{8 \times 7 \times 6}{3 \times 2 \times 1} = 56$.

10. C The first digit can be 1-9; the second, 0-9; the third can only be 0, 2, 4, 6, 8 or 5 digits. Then there are $9 \times 10 \times 5$ or 450 even 3-digit numbers.

Number Bases

Our number system is called base 10 because it is composed of ten digits; the numbers 0 to 9. A multi-digit number represents powers of 10.

EXAMPLE: The number 584 represents $(5 \times 10^2) + (8 \times 10^1) + (4 \times 10^0)$.

Other number bases are often used in mathematics. The computer is really a base 2 machine. The base 2 is composed of the numbers 0 and 1. A base 5 number system uses the numbers 0, 1, 2, 3, 4. Some of the CLEP examination problems might ask you to represent a base 5 number, for example, in base ten.

EXAMPLE: The number 234_5 (read two-three-four base five) represents:

$$= (2 \cdot 5^2) + (3 \cdot 5^1) + (4 \cdot 5^0)$$
$$= 2(25) + 3(5) + 4(1)$$
$$= 50 + 15 + 4$$
$$= 69 \text{ in base 10}$$

The number 10101_2 could be changed to its base 10 equivalent in the same way:

EXAMPLE: The base 10 equivalent of 10101_2

$$= (1 \cdot 2^4) + (0 \cdot 2^3) + (1 \cdot 2^2) + (0 \cdot 2^1) + (1 \cdot 2^0)$$
$$= 16 + 0 + 4 + 0 + 1$$
$$= 21$$

To change a base 10 number into another base equivalent, divide by powers of the base in descending order.

EXAMPLE: Change 134 into its base 6 equivalent.

SOLUTION: Divide by the largest power of 6 possible:

$6^2 = 36$ then: $36\overline{)134}$ gives 3   There are 3 powers of 6^2.
         $\underline{108}$
          26

And: $6^1 = 6$ Then: $6\overline{)26}$ gives 4   There are 4 powers of 6^1.
            $\underline{24}$
             2

There are 2 powers of 6^0.

124

Therefore, 134_{10} is the equivalent of 342_6. To prove, change back to base 10. Thus 342_6 is equivalent to:

$3(6^2) + 4(6^1) + 2(6^0) = 3(36) + 4(6) + 2(1) = 108 + 24 + 2 = 134.$

Logarithms

A **log** is an exponent of a number. For example, to say $2^3 = 8$, we can write $\log_2 8 = 3$, which is read "the log of eight base two is three." Logs can be of any number base. $\text{Log}_a x = b$ is the same thing as saying, in exponential form, $a^b = x$. The common logarithms are to the base 10. Thus, since $10^2 = 100$, $\log_{10} 100 = 2$. If the base is not written, it is understood that the base is 10. So, $\log 14 = x$ says $10^x = 14$. You can see that any number between powers of 10 will have a log that is part decimal.

$10^1 = 10$, therefore $\log 10 = 1$

$10^2 = 100$, therefore $\log 100 = 2$

$10^3 = 1000$, therefore $\log 1000 = 3$

$10^? = 25$, therefore $\log 25$ = a number between 1 and 2

Logarithms of numbers can be found in tables in most mathematics books that deal with the topic. For purposes of the exam you need only approximate. Most of the examination questions deal with the properties of logs in general rather than specific numbers. Logs are used to facilitate computation with large and small numbers. The properties of logs that make them a useful tool are as follows:

Log AB = Log A + Log B

Log A ÷ B = Log A - Log B

Log A^b = b Log A

EXAMPLE: To compute $\dfrac{5,234 \times (3,758)^2}{2,714}$, we could find the log of 5,234, add to it twice the log of 3,758, and subtract the log of 2,714. This gives us the log of the answer, called the anti-log. Using the log table we can find the number which has that log. In doing the log problems on the examination it is important to remember that logarithms are really exponents.

PRACTICE PROBLEMS — LOGARITHMS

1. Represent the following in exponential form:

    A. $\log_2 4 = 2$

    B. $\log_3 81 = 4$

    C. $2 = \log_4 16$

    D. $\log_9 3 = \frac{1}{2}$

    E. $\log_b A = x$

2. Represent the following in log form:

    A. $2^3 = 8$

    B. $3^3 = 27$

    C. $4^{\frac{1}{2}} = 2$

    D. $A^x = B$

    E. $8^{\frac{1}{3}} = 2$

3. Find the value of x:

    A. $\log_x 216 = 3$

    B. $\log_x 81 = 4$

4. Show the following are true:

    A. $\log_2 64 = 3 \log_8 64$

    B. $\log_2 8 \cdot \log_8 2 = 1$

    C. $\log_3 27 \cdot \log_{27} 3 = 1$

5. If log 758 = 2.8797 and log .416 = (9.6191 - 10), find log (758 · .416).

6. If log 2 = .3010, find log of 2^5.

ANSWERS:

1.   Log Form: Exponential Form:

    A. $\log_2 4 = 2$ $2^2 = 4$

    B. $\log_3 81 = 4$ $3^4 = 81$

    C. $2 = \log_4 16$ $4^2 = 16$

    D. $\log_9 3 = \frac{1}{2}$ $9^{\frac{1}{2}} = 3$ (Remember $9^{\frac{1}{2}} = \sqrt{9} = 3$)

    E. $\log_b A = x$ $b^x = A$

2.   Exponential Form: Log Form:

    A. $2^3 = 8$ $\text{Log}_2 8 = 3$

    B. $3^3 = 27$ $\text{Log}_3 27 = 3$

    C. $4^{\frac{1}{2}} = 2$ $\text{Log}_4 2 = \frac{1}{2}$

    D. $A^x = B$ $\text{Log}_A B = X$

    E. $8^{\frac{1}{3}} = 2$ $\text{Log}_8 2 = \frac{1}{3}$

3. A. $\log_x 216 = 3$, then $x^3 = 216$ and $x = \sqrt[3]{216} = 6$ (because $6 \cdot 6 \cdot 6 = 216$)

    B. $\log_x 81 = 4$, then $x^4 = 81$ and $x = \sqrt[4]{81} = 3$
(because $3 \cdot 3 \cdot 3 \cdot 3 = 81$)

4. Show the following are true:

    A. $\text{Log}_2 64 = x$ $3\,\text{Log}_8 64 = x$

            $2^x = 64$ $8^x = 64$

             $x = 6$ $x = 2$

    $(2 \cdot 2 \cdot 2 \cdot 2 \cdot 2 \cdot 2 = 64)$ $3(2) = 6$

B. $\text{Log}_2 8 \cdot \text{Log}_8 2 = 1$

$2^x = 8$ $8^x = 2$

$x = 3$ $x = \frac{1}{3}$

$$3 \cdot \frac{1}{3} = 1$$

C. $\text{Log}_3 27 \cdot \text{Log}_{27} 3 = 1$

$3^x = 27$ $27^x = 3$

$x = 3$ $x = \frac{1}{3}$

$$3 \cdot \frac{1}{3} = 1$$

NOTE: This example is true for all problems where the base and number are switched. In general, $\text{Log}_b A \cdot \text{Log}_a B$ will always equal 1

(i.e., $\text{Log}_3 12 \cdot \text{Log}_{12} 3 = 1$).

5. NOTE: In this example the value of the logs was obtained from a log table.

Using the Law of Logs which says Log AB = Log A + Log B:

Log $(758 \cdot .416)$ = 2.8797 + (9.6191 - 10)

= 2.8797

(9.6191 - 10)
—————————————
12.4988 - 10

= 2.4988

If you look up the number .4988 in a log table, you will find it represents the number 3153. With a characteristic of two, this would represent the number 315.3. Multiply the numbers 758 × .416 to prove this is correct (approximately).

6. Using the Law of Logs:

Log a^b = b log a

Log 2^5 = 5 log 2

= 5(.3010)

= 1.5050

If you look up .5050 in a log table, you will find that it represents the number 32. Does $2^5 = 32$?

Imaginary and Complex Numbers

Since there is no real number to express the square root of a negative number, mathematicians use the letter "i" to represent $\sqrt{-1}$. The number i is called the imaginary unit. By using i we can express $\sqrt{-8}$, for example, as $i\sqrt{8}$. Thus $\sqrt{-16}$ could be represented by 4i. By using i we can perform operations with real and imaginary numbers. To facilitate operations with imaginary numbers it is helpful to see the effect of exponents on i.

$$i^1 = i = \text{definition}$$

$$i^2 = -1 = \text{definition}$$

$$i^3 = i^2 \cdot i = -1(i) = -i$$

$$i^4 = i^2 \cdot i^2 = -1 \cdot -1 = 1$$

Each succeeding power will repeat the pattern i, -1, -i and 1.

The examination may require you to simplify expressions involving imaginary units.

EXAMPLE: $(3 + \sqrt{-5})(3 - \sqrt{-5}) = 3^2 - (\sqrt{-5})^2$

$$= 3^2 - (i\sqrt{5})^2 = 3^2 - i^2(\sqrt{5})^2$$

$$= 9 - 5i^2$$

$$= 9 - 5(-1)$$

$$= 9 + 5 = 14$$

Numbers composed of a real number and an imaginary part are known as complex numbers. The general form of a complex number is a + bi, where a is the real part and b the imaginary part of a + bi, a and b being real numbers. To perform operations with complex numbers think of i as a variable and change any expression with powers of i by using the relationships shown above.

EXAMPLE: $(3 + 2i)(7 + 4i) = (3)(7) + (3)(4i) + (7)(2i) + (2i)(4i)$

$$= 21 + 12i + 14i + 8i^2$$

$$= 21 + 26i + 8(-1)$$

$$= 21 + 26i - 8$$

$$= 13 + 26I$$

PRACTICE PROBLEMS — IMAGINARY AND COMPLEX NUMBERS

1. Multiply:

    A. $2(5i)$

    B. $(7i)(6i)$

    C. $(-15i)(\frac{1}{3}i)$

    D. $\sqrt{-9} \cdot \sqrt{-3}$

2. Add:

    A. $\sqrt{-16} + \sqrt{-49}$

    B. $5\sqrt{-48} - \frac{1}{3}\sqrt{-27}$

    C. $5\sqrt{-2} + \sqrt{-8}$

3. Simplify:

    A. i^9

    B i^{14}

    C. $\dfrac{8\sqrt{-32}}{2\sqrt{-2}}$

    D. $(2 + 3i)(3 - 2i)$

    E. $(3 - i)(4 + i)$

ANSWERS:

1. Multiply:

    A. $2(5i) = 10i$

    B. $(7i)(6i) = 42i^2 = 42(-1) = -42$

    C. $(-15i)(\frac{1}{3}i) = -5i^2 = -5(-1) = 5$

    D. $\sqrt{-9} \cdot \sqrt{-3}\ =\ 3i\ i\sqrt{3}$

    $$= 3i^2\sqrt{3}$$
    $$= 3(-1)\sqrt{3}$$
    $$= -3\sqrt{3}$$

2. Add:

A. $\sqrt{-16} + \sqrt{-49}$ = 4i + 7i = 11i

B. $5\sqrt{-48} - \dfrac{1}{3}\sqrt{-27}$ = $5i\sqrt{48} - \dfrac{1}{3}i\sqrt{27}$

$$= 5i\sqrt{16}\cdot\sqrt{3} - \frac{1}{3}i\sqrt{9}\cdot\sqrt{3}$$

$$= 20i\sqrt{3} - i\sqrt{3}$$

$$= 19i\sqrt{3}$$

C. $5\sqrt{-2} + \sqrt{-8}$ = $5i\sqrt{2} + i\sqrt{8}$

$$= 5i\sqrt{2} + i\sqrt{4}\sqrt{2}$$

$$= 5i\sqrt{2} + 2i\sqrt{2}$$

$$= 7i\sqrt{2}$$

3. To simplify these examples use the following:

$$i^1 = i = \text{definition}$$

$$i^2 = -1 = \text{definition}$$

$$i^3 = (i^2)\, i = -1(i) = -i$$

$$i^4 = (i^2)(i^2) = (-1)(-1) = 1$$

A. $i^9 = (i^4)(i^4)(i) = (1)(1)(i) = i$

B $i^{14} = (i^4)(i^4)(i^4)(i^2) = (1)(1)(1)(-1) = -1$

C. $\dfrac{8\sqrt{-32}}{2\sqrt{-2}} = \dfrac{8i\sqrt{32}}{2i\sqrt{2}} = \dfrac{8i\sqrt{16}\sqrt{2}}{2i\sqrt{2}} = \dfrac{32i\sqrt{2}}{2i\sqrt{2}} = 16$

D. (2 + 3i)(3 - 2i) = (2)(3) + (2)(-2i) + (3i)(3) + (3i)(-2i)

$$= 6 + (-4i) + 9i + (-6i^2)$$

$$= 6 - 4i + 9i + (-6)(-1)$$

$$= 6 + 5i + 6 = 12 + 5i$$

E. (3 - i)(4 + i) = (3)(4) + (3)(i) + (-i)(4) + (-i)(i)

$$= 12 + 3i - 4i - i^2$$

$$= 12 - i - (-1)$$

$$= 13 - i$$

131

Taking the CLEP Practice Examination in Mathematics

Now that you have finished the entire review section, you are ready to take a practice test. The test follows the format of the CLEP General Examination in Mathematics.

It is important that you take this test now. While you may have done well on each type of question in isolation in the preceding section of the book, another skill is necessary. You must practice working with a variety of questions. In effect, you need to practice "switching gears" or training your mind to go from one type of question to another in a short period of time.

For best results try to simulate the test situation as nearly as possible. The following procedure should be followed for maximum benefit:

1. Find a quiet spot where you won't be disturbed.

2. Time yourself accurately. Work 45 minutes in Part A and 45 minutes in Part B. Don't quit until the time is up!

3. Use the coding system for a systematic approach to the examination.

After you finish:

1. Check your answers.

2. Review procedures for doing questions if a particular type is giving you trouble.

Warm-up Exercise

Before you get started with the actual practice exam in the next section, do the 15 problems in the warm-up exercise. This is not a timed exercise. However, if you struggle through these problems, you should consider additional review before attempting the Practice Examination.

1. A person can choose any 2 of 8 items from menu list I and any 3 of 7 items from menu list II. Approximately how many different meal combinations are possible?

   (A) 1,000
   (B) 500
   (C) 1,500
   (D) 750

2. In the conditional p --> q, p is true and q is false. What can be said about q → p?

   (A) q → p is false
   (B) q → p is true
   (C) p → q is true
   (D) p ∧ q is true

3. If for real numbers (x,y), $f(x) = x^2 + 3x - 1$, $f(x - 2) =$

   (A) $x^2 + 3x - 2$
   (B) $x^2 - x + 3$
   (C) $x^2 - x - 3$
   (D) $2x^2 - 3x + 3$

4. Which of the following operations are commutative?

   I. Addition
   II. Subtraction
   III. Multiplication
   IV. Division

   (A) I and II
   (B) I and III
   (C) II and IV
   (D) I, II, III, and IV

5. If $\sqrt{-1}$ is represented by i, what is $\sqrt{-8} \times \sqrt{-2}$?

   (A) 4i
   (B) 2i
   (C) -4
   (D) 16i

6. The equation of the graph shown below could only represent which of the following equations?

   (A) y = 3x - 4
   (B) y = x² - x
   (C) y = x³ - x
   (D) 3y = x - 4

7. If A = {b, c, d} and B = {6, 7}, how many elements are in A x B?

   (A) 3
   (B) 6
   (C) 5
   (D) 2

8. Change 132_5 to base 10 notation:

   (A) 52
   (B) 48
   (C) 65
   (D) 42

9. A person starts a chain letter by writing two friends and requesting that each send a copy to two others. If the chain is unbroken after the fourth set is mailed, what is the total spent for postage at $.32 a letter?

   (A) $9.92
   (B) $9.60
   (C) $5.12
   (D) $8.96

10. Five students receive test scores of 62, 84, 99, 77, and 88. What is the average of the mean and the median scores?

    (A) 82
    (B) 83
    (C) 85
    (D) 86

11. Which of the following is equivalent to $\sqrt{\sqrt{a^{12}}}$?

    (A) a^3
    (B) a^4
    (C) a^2
    (D) a^6

12. A bicycle lock has a combination consisting of three numbers. If only the numbers 0 through 5 are on the dial, how many possible combinations are there?

    (A) 216
    (B) 15
    (C) 125
    (D) 16

13. For the function $y = 2x + 1$, $x = \{0, 1, 2\}$, what is the domain of the inverse of the function?

    (A) {1, 3, 5}
    (B) {0, 1, 5}
    (C) {0, 1, 3}
    (D) {0, 1, 2}

14. The line, which passes through the point (4, 6) with a slope of 2, has which of the following for its equation?

    (A) $y = 2x + 2$
    (B) $y = 2x - 2$
    (C) $y = 2x$
    (D) $y = \frac{1}{2}x - 2$

15. For what value(s) of x is the domain restricted if the function $f(x) = \dfrac{x^2 - y^2}{x - y}$?

(A) None
(B) $x \neq 0$
(C) $x \neq y$
(D) $x \neq -y$

WARM-UP SOLUTIONS

1. A There are $_8C_2 = \dfrac{8 \cdot 7}{2 \cdot 1} = 28$ possible combinations from list I, and

$_7C_3 = \dfrac{7 \cdot 6 \cdot 5}{3 \cdot 2 \cdot 1} = 35$ possible combinations from list II. Altogether there are $28 \cdot 35 = 980$ or approximately 1,000 possible choices.

2. B The truth table for a conditional shows that when the antecedent is false but the consequent is true the conditional is considered true.

3. C $f(x - 2) = (x - 2)^2 + 3(x - 2) - 1 = (x^2 - 4x + 4) + 3x - 6 - 1 = x^2 - x - 3$.

4. B Addition $(a + b = b + a)$ and multiplication $(ab = ba)$ are both commutative.

Subtraction $(b - a \neq a - b)$ and division $(b \div a \neq a \div b)$ are not commutative.

5. C $\sqrt{-8} \cdot \sqrt{-2} = i\sqrt{8} \cdot i\sqrt{2} = i^2 \cdot \sqrt{8 \cdot 2} = -1 \cdot \sqrt{16} = -4$

6   C Choices A and B are linear equations and the graph of a linear equation is a straight line. Choice B is a parabola. Choice C is the only possible answer.

7. B A X B is the symbol which asks for the Cartesian Product of sets A and B. In general, if Set A has x elements and Set B has y elements, A X B has xy elements.

8. D $132_5 = 1 \cdot 5^2 + 3 \cdot 5^1 + 2 \cdot 5^0 = 25 + 15 + 2 = 42$

9. B After the first set is mailed, there are $2^1 = 2$ letters. After the second set, there are $2^2 = 4$ more. After the third, there are $2^3 = 8$ more. After the fourth, there are $2^4 = 16$ more. $16 + 8 + 4 + 2 = 30$ letters at \$0.32 each = \$9.60.

10. B The mean, or arithmetic average is $\dfrac{62 + 84 + 99 + 77 + 88}{5} = 82$.

The median is the middle score of the numbers arranged in order 62, 77, 84, 88, 99. since there are five numbers, the middle score is the third number, or 84. The average of 82 and 84 is 83.

11. A $\sqrt{\sqrt{a^{12}}} = \sqrt{a^6} = a^3$. Note: the square root of a variable with an even exponent is the variable with the exponent divided by 2. For example $\sqrt{a^{12}} = a^6$, and $\sqrt{a^6} = a^3$.

12. A Since each dial has 6 possible numbers to choose, there will be $6 \cdot 6 \cdot 6$ possible combinations of lock numbers.

13. A   The domain of the inverse is the same as the range of the given function. For $y = 2x + 1$, the range would be $2(0) + 1 = 1$, $2(1) + 1 = 3$, and $2(2) + 1 = 5$ or $\{1, 3, 5\}$.

14. B   The general equation of a straight line is $y = mx + b$, where m = slope and b = y-intercept. Since we know the slope = 2, our equation is in the form $y = 2x + b$. Since the point passes through the point (4, 6), the values 4 and 6 must satisfy the equation. So, $6 = 2(4) + b$ and b, therefore, is -2. The equation is $y = 2x - 2$.

15. C   If $x = y$, then the denominator is equal to zero and division by zero is impossible.

PART A (45 MINUTES)

1. The product of $(x + y)(x - y) =$

   (A) $x^2 + 2xy + y^2$
   (B) $x^2 - y^2$
   (C) $x^2 + 2xy - y^2$
   (D) $x^2 - 2xy - y^2$

2. Colleen has a deck of cards and a single die. If she picks a card and rolls the die, what is the probability that the card is a spade and the die shows an even number?

   (A) 3/4
   (B) 1/8
   (C) 1/4
   (D) 1/2

3. The $\sqrt{28}$ is equal to

   (A) 5
   (B) $7\sqrt{4}$
   (C) $4\sqrt{7}$
   (D) $2\sqrt{7}$

4. In the Venn Diagram, which relationship does the shaded region represent?

   (A) $(A \cap B) \cup C$
   (B) $A \cap (B \cup C)$
   (C) $A \cup (B \cup C)$
   (D) $(A \cup B) \cap C$

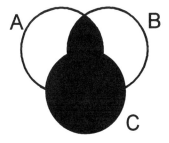

5. If $2x - 5 = 7$, then $x =$

   (A) 2
   (B) 12
   (C) 6
   (D) 1

6. What was the original price of a watch that sold at a 25% discount for $45.00

   (A) $60.00
   (B) $11.25
   (C) $56.25
   (D) $75.00

7. Which of the following is not a factor of $x^4 - 81$

   (A) $(x^2 + 9)$
   (B) $(x - 3)$
   (C) $(x + 3)$
   (D) $(x + 9)$

8. Sandi's scores on her math exams this semester are 93, 82, 87, 78, and 90. Her median score was

   (A) 93
   (B) 86
   (C) 87
   (D) 78

9. If $f(X) = \dfrac{3X + 5}{6X}$, what is f(5)?

   (A) 2/3
   (B) 20
   (C) 2
   (D) 4

10. What is the solution set for $\left|\dfrac{2}{5}x - 3\right| = 11$?

    (A) {35}
    (B) {20}
    (C) {20, 35}
    (D) {-20, 35}

11. Which of the following is irrational?

(A) $\sqrt{81}$

(B) $\sqrt{7}$

(C) $\dfrac{\sqrt{49}}{16}$

(D) $\dfrac{3}{\sqrt{36}}$

12. How many prime numbers are there between 35 and 50?

(A) 3
(B) 4
(C) 5
(D) 6

13. -54 - (-28) =

(A) -26
(B) -82
(C) 26
(D) 82

14. If f(X) = (X² + 4), and g(Y) = (Y-5), what is f[g(4)]?

(A) 85
(B) 18
(C) 5
(D) 77

15. If $f(X) = X^2 - X + 7$, then f(-6) =

(A) 37
(B) -23
(C) -35
(D) 49

16. The conditional statement "If Marissa isn't tired, she can stay up another hour" is logically equivalent to which of the following?

(A) If she isn't tired, Marissa can't stay up another hour.
(B) If Marissa is tired, she can't stay up another hour.
(C) If she can stay up another hour, Marissa isn't tired.
(D) If she can't stay up another hour, Marissa is tired.

17. The sum of $(3x^2 - 7x + 9)$ and $(5x^2 + 12x - 13)$ is

   (A) $8X^2 + 5X - 4$
   (B) $8X^2 + 19X - 4$
   (C) $8X^2 - 5X - 4$
   (D) $8X^2 + 5X + 22$

18. $\dfrac{3}{\sqrt{12}}$ is equal to

   (A) $\dfrac{\sqrt{3}}{2}$
   (B) $\dfrac{3}{2}$
   (C) $\dfrac{\sqrt{3}}{6}$
   (D) $\dfrac{3\sqrt{3}}{2}$

19. Which product is larger, $(-7)(5)$, $(-7)(-5)$, or $(7)(-5)$?

   (A) $(-7)(5)$
   (B) $(-7)(-5)$
   (C) $(7)(-5)$
   (D) they're all equal

20. The product of $(X - 8)$ and $(X + 5)$ is

   (A) $X^2 + 13X - 40$
   (B) $X^2 + 3X - 40$
   (C) $X^2 - 3X + 40$
   (D) $X^2 - 3X - 40$

21. What are the factors of $8X^5 - 12X^{25}$?

   (A) $4X^5(2 - 3X^5)$
   (B) $8X^5(X - 4X^5)$
   (C) $4X^5(2 - 3X^{20})$
   (D) $8X^5(1 - 4X^{20})$

22. What is the product of 3 consecutive integers whose sum is 27?

(A) 729
(B) 720
(C) 990
(D) 504

23. If X represents an even number and Y represents an odd number, which of the following is(are) not odd?

I. $X + Y + 10$

II. $3Y + 5$

III. $XY + X$

(A) I only
(B) II only
(C) I & II
(D) II & III

24. Find the product of 3^4 and 9^3.

(A) 3^7
(B) 3^{12}
(C) 3^{10}
(D) 9^7

25. Simplify $\sqrt{75}$.

(A) $5\sqrt{3}$
(B) $2\sqrt{3}$
(C) $3\sqrt{2}$
(D) $3\sqrt{5}$

26. What is the solution set for $X^2 - 11X = -24$?

(A) {6,4}
(B) {-3, -8}
(C) {3, 8}
(D) {-12, 2}

Use the graph, below in question 27, to solve problems 27 through 29.

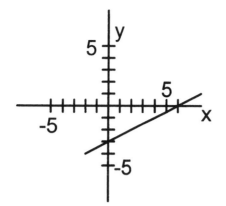

27. What is the y-intercept of the line that passes through the given points?

(A) (0, -3)
(B) (6, 0)
(C) (-6, 0)
(D) (0, 3)

28. What is the slope of the line that passes through the given points?

(A) 2
(B) 1/2
(C) -2
(D) -1/2

29. What is the equation of the line that passes through the given points?

(A) y = 1/2 X + 3
(B) y = 1/2X - 3
(C) y = -3X + 1/2
(D) y = 2X - 3

30. If A = {positive even integers < 25} and B = {multiples of 5 < 30}, then how many elements are there in A ∩ B?

(A) 3
(B) 17
(C) 2
(D) 5

144

31. In a recent poll of moviegoers, 7 out of 10 women said they would recommend the movie to a friend, but only 4 out of 10 men agreed. Based on this information, what is the probability that the person is a man, given that the person would not recommend the movie to a friend?

(A) 2/3
(B) 1/3
(C) 6/10
(D) 4/11

32. Which is larger, $|(-3)(8)|$, $(-2)(-12)$, or $(4)(6)$?

(A) $|(-3)(8)|$
(B) $(-2)(-12)$
(C) $(4)(6)$
(D) They're all equal

PART B (45 MINUTES)

33. If $A = \{1,2,3,4\}$, $B = \{2,3,5,6\}$, what is $A \cup B$?

(A) $\{2,3\}$
(B) $\{1,2,5,6\}$
(C) $\{1,2,3,4,5,6\}$
(D) $\{2\}$

34. Which of the following is not ALWAYS true?

I. $\sqrt{a} \times \sqrt{b} = \sqrt{ab}$
II. $\sqrt{a} \div \sqrt{b} = \sqrt{a \div b}$
III. $\sqrt{a} + \sqrt{b} = \sqrt{a + b}$

(A) I only
(B) I, II only
(C) III only
(D) II only

35. Which of the following could be the equation of the graph shown?

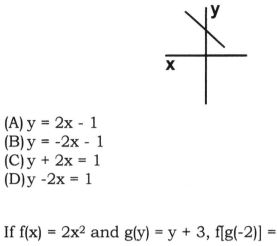

(A) y = 2x - 1
(B) y = -2x - 1
(C) y + 2x = 1
(D) y -2x = 1

36. If f(x) = 2x² and g(y) = y + 3, f[g(-2)] =

(A) -2
(B) 2
(C) -8
(D) 1

37. If p represents the statement "It is cold", and q represents the statement "I will buy a coat", which of the following represents the statement "If it is cold, then I will buy a coat"?

(A) ~p → ~q
(B) p → ~q
(C) p → q
(D) q → p

38. What is the solution set for |x - 2| = 3?

(A) {5, -1}
(B) {5}
(C) {1, 5}
(D) {-5, 1}

39. How many subsets of the set {a, b, c, d, e} are there?

(A) 5
(B) 10
(C) 25
(D) 32

40. If you draw a card and replace it, and then draw another card, what is the probability that both cards are spades?

   (A) 3/13
   (B) 1/16
   (C) 1/169
   (D) 1/2

41. If $a*b = a^2+b^2$, what is $(2*1)*3$?

   (A) 34
   (B) 18
   (C) 13
   (D) 15

42. If $f(x) = \sqrt{25 - x^2}$ is to be a real number, the domain of x must be:

   (A) $x > 5$
   (B) $x < 5$
   (C) $-5 \leq x \leq 5$
   (D) $-5 \geq x \geq 5$

43. Which of the following sets is not closed under multiplication?

   (A) {Even numbers}
   (B) {Positive numbers}
   (C) {Negative numbers}
   (D) {Integers}

44. How many prime numbers are there between 50 and 60?

   (A) 1
   (B) 2
   (C) 3
   (D) 4

45. If A = {2,3,4}, B = {1,2,3}, and C = {1,5,8}, then (A ∩ B) ∪ C =

(A) {1,2,3,4,5,8}
(B) {2,3}
(C) ∅ or { }
(D) {1,2,3,5,8}

46. The number 4 is equivalent to which of the following base 2 numbers?

(A) 100_2
(B) 11_2
(C) 10_2
(D) 101_2

47. If $a = b^3$, when b is doubled then a is

(A) tripled
(B) doubled
(C) multiplied by 8
(D) multiplied by 6

48. If a is an odd integer, which of the following must be even?

(A) 3a
(B) 3a + 2
(C) 2a - 1
(D) 3a + 1

49. If $x^2 - 3x = 28$ then x =

(A) {7, -4}
(B) {4, 7}
(C) {-7, 4}
(D) {-7, -4}

50. How many different groups of 3 people can be formed from a group of 10?

(A) 100
(B) 120
(C) 720
(D) 220

51. How many common solutions do f(x) = x² and f(x) = 2x + 3 have in common?

(A) 1
(B) 2
(C) 0
(D) 3

52. If A = {x: x ≥ 1}, and B = {x: x ≤ 1}, then A ∩ B =

(A) ∅ or { }
(B) {1}
(C) {-1}
(D) { All Integers}

53. What is the next number in the chart below?

x	0	1	2	3	4
y	0	0	2	6	?

(A) 8
(B) 12
(C) 16
(D) 12

54. If $\sqrt{ab} = 6$, and a and b are integers, which of the following could **NOT** be the value of a – b?

(A) 0
(B) -35
(C) -16
(D) 12

55. If the conditional statement "If a = 2, then b = 3" is true, which of the following is also true?

(A) If a ≠ 2, then b ≠ 3
(B) If b = 3, then a = 2
(C) If b ≠ 3, then a ≠ 2
(D) If b = 3, then a ≠ 2

56. The shaded part of the Venn diagram shown below represents which of the following?

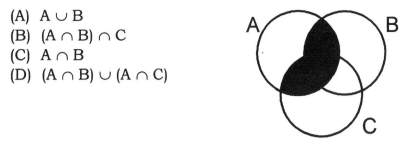

    (A) $A \cup B$
    (B) $(A \cap B) \cap C$
    (C) $A \cap B$
    (D) $(A \cap B) \cup (A \cap C)$

57. In the formula $P = 2\pi \dfrac{\sqrt{L}}{g}$, if g is a constant, what must L be multiplied by so P will be tripled?

    (A) 8
    (B) 3
    (C) 4
    (D) 9

58. Which of the following properties of logarithms are true?

    I. $\text{Log } AB = \text{Log } A + \text{Log } B$

    II. $\text{Log } (A \div B) = \text{Log } A - \text{Log } B$

    III. $\text{Log } A^b = b \text{ Log } A$

    (A) I only
    (B) I and III
    (C) I and II
    (D) I, II, and III

59. Which of the following is always true for real numbers (x and y > 0)?

    I. $(x + y)^a = x^a + y^a$

    II. $(x^a)(y^a) = (xy)^a$

    III. $x^a \div y^a = (x \div y)^a$

    (A) I and III
    (B) II and III
    (C) I, II, and III
    (D) None

60. If x is a real number, $x^2 < x$ is true for:

(A) x > 1
(B) x < 1
(C) x = 0
(D) 0 < x < 1

61. The statement ~a → b is false if:

(A) a is true
(B) b is true
(C) b is false and a is true
(D) a is false and b is false

62. If $\sqrt{-1}$ is represented by i, what is (4 - 2i)(4 + 2i)?

(A) 20
(B) 16
(C) 16 - 4i
(D) 20i

63. The graph of the straight line y = f(x) passes through the point (3, 5) and also through (4, a). If the slope of the line is 3, what is the value of a?

(A) 6
(B) 12
(C) 14
(D) 8

64. A box contains cards numbered 1 through 21. What is the probability that the first card drawn is divisible by 2 and the second card drawn is divisible by 5 if the first card is replaced after drawing?

(A) 40/441
(B) 14/21
(C) 12/21
(D) 44/441

65. If $\log_{10} 5 = b$ and $\log_{10} 7 = c$, then $\log_{10} 35 = ?$

(A) b + c
(B) b - c
(C) bc
(D) b/c

Answers to Practice Examination

PART A

1. B   x multiplies both x and (-y), and (+y) multiplies both x and (-y). The result is $x^2 + xy - xy - y^2$. The middle terms cancel each other out, and the answer is $x^2 - y^2$.

2. B   These are independent events, so the answer is the product of the individual probabilities.

$$P(\text{spade}) = \frac{1}{4}$$

$$P(\text{even}) = \frac{1}{2}$$

$$P(\text{spade and even}) = \frac{1}{4} \cdot \frac{1}{2} = \frac{1}{8}$$

3. D $\sqrt{28} = \sqrt{4}\sqrt{7} = 2\sqrt{7}$

4. A   The solution is the area that is in both circles A and B plus the area enclosed by circle C.

5. C
$$2x - 5 = 7$$
$$\underline{+5 = 5}$$
$$2x + 0 = 12$$

$$x = \frac{12}{2} = 6$$

6. A   If the discount is 25%, then $45.00 represents 75% of the original cost.

$$.75x = 45$$
$$x = 45/.75 = 60$$

152

7. D $x^4 - 81$ is the difference between 2 squares. Initially, we write it as the product of $(x^2 - 9)(x^2 + 9)$. $(x^2 - 9)$ is also the difference of 2 squares whose factors are $(x - 3)(x + 3)$. The 3 factors are $(x - 3)(x + 3)(x^2 + 9)$. The term $(x + 9)$ is not a factor.

8. C   The median score is the number for which there are as many scores below it as above it. If you put the scores in order, you can see that the middle score is 87. 86 is the mean.

9. A $\dfrac{3(5) + 5}{6(5)} = \dfrac{15 + 5}{30} = \dfrac{20}{30} = \dfrac{2}{3}$

10. D

2/5 x - 3 = 11	and	2/5 x - 3 = -11
2/5 x = 14	and	2/5 x = -8
x = 35	and	x = -20

11. B $\sqrt{81} = 9$, $\dfrac{\sqrt{49}}{16} = \dfrac{7}{16}$, and $\dfrac{3}{\sqrt{36}} = \dfrac{3}{6} = \dfrac{1}{2}$. $\sqrt{7}$ is the only one that cannot be expressed as a fraction of integers.

12. B   The prime numbers are 37, 41, 43, and 47.

13. A   -54 - (-28) = -54 + 28 = -26

14. C   g(4) = 4 - 5 = -1. f (-1) = [(-1)² + 4] = 1 + 4 = 5

15. D   f (-6)   =   (-6)² - (-6) + 7
                  = 36 + 6 + 7
                  = 49

16. D   If A = tired and B = stay up, then the statement can be expressed as ~A → B. The contrapositive is logically equivalent, which in this case would be ~B → A, or choice D.

17. A   Just add like terms together, combining the coefficients using our rules for signed numbers.

18. A   Multiply the numerator and denominator by $\sqrt{3}$. The result is $\dfrac{3\sqrt{3}}{\sqrt{36}} = \dfrac{3\sqrt{3}}{6} = \dfrac{\sqrt{3}}{2}$.

19. B   The first and third products are negative, the second product is positive.

20. D   x multiplies both x and (+5), and (-8) multiplies both x and (+5). The result is $x^2 + 5x - 8x - 40$. Combining the middle terms, we get $x^2 - 3x - 40$.

21. C   4 goes evenly into both 8 and 12. x^5 goes evenly into both x^5 and x^{25}. The common factor is $4x^5$. Remember, when you divide x^5 into x^{25}, you subtract the exponents.

22. B   The easiest way to solve this problem is to divide 27 by 3, which equals 9, and add and subtract 1 from 9. The three numbers are 8, 9, and 10. Their product is 720.

23. D   Pick any even number for X, say 2, and any odd number for Y, say 1. Then:
I = 2 - 1 + 10 = 13
II = 3(1) + 5 = 8
III = (2)(1) + 2 = 4.

II and III are not odd.

24. C   Before you combine the exponents, you must have a common base, so you must rewrite $\quad 9^3 : 9^3 = (3^2)^3 = 3^{2 \times 3} = 3^6$. Then just add the exponents of 3 together.

25. A $\sqrt{75} = \sqrt{25}\sqrt{3} = 5\sqrt{3}$.

26. C   Add 24 to both sides so that the equation is equal to 0. Then write in factored form (x - 8) (x - 3) = 0. x - 8 = 0 or x - 3 = 0. Solving for x, x = 8 or 3.

27. A   The y-intercept is the point where the x value is 0, so the only possible answers are (A) or (D). Since the line crosses the y-axis below the x-axis, we know that the y-intercept is negative.

28. B   The line is rising from left to right, so the slope is positive. The slope is a slow rise, so it's probably a fraction less than 1. To calculate the slope, we take the difference in the y coordinates over the difference in the x coordinates.

$$\frac{(-3 - 0)}{(0 - 6)} = \frac{-3}{-6} = \frac{1}{2}$$

29. B $y = mx + b$; $m = \frac{1}{2}$ and $b = -3$.

30. C   The intersection is {10, 20}.

31. A   There are 6 men who didn't like the movie and 3 women, for a total of 9 people. The probability that the person was a man is $\frac{6}{9}$ or $\frac{2}{3}$.

32. D   In each case the result is a positive 24.

PART B

33. C   The union contains any element that is in either set.
         $A \cup B = \{1, 2, 3, 4, 5, 6\}$

34. C   Property III is generally not true. Take, for example, a = 4 and b = 9. Then $\sqrt{4} + \sqrt{9} = 2 + 3 = 5$, which is not equal to $\sqrt{4 + 9} = \sqrt{13}$.

35. C We can see by inspection that the line has a negative slope and a positive y-intercept. Writing answer C in the form y = mx + b, we get y = -2x + 1. It is the only response that has a negative slope and a positive y-intercept.

36. B g(-2) = -2 + 3 = 1. F(1) = 2(1)² = 2.

37. C The "if-then" statement is called the conditional and its logic symbol is →. So, "if p, then q" becomes p → q.

38 A |x - 2| is true if either x - 2 = 3 or x - 2 = -3. x - 2 = 3 when x = 5 and x - 2 = -3 when x = -1.

39. D The number of subsets of a set with n elements is given by 2^n. Since this set has 5 elements, there are 2^n = 32 subsets. (Note: this includes the null set and the set itself.)

40. B The Probability that two independent events will happen is found by multiplying their individual probabilities together:

$$P(\text{1st Spade}) = \frac{1}{4}; \quad P(\text{2nd Spade}) = \frac{1}{4}; \quad P(\text{drawing 2 Spades}) = \frac{1}{4} \cdot \frac{1}{4} = \frac{1}{16}.$$

41. A If a * b = a² + b², then

$$(2 * 1) * 3 = (2^2 + 1^2) * 3$$
$$= (4 + 1) * 3$$
$$= 5 * 3$$
$$= 5^2 + 3^2$$
$$= 25 + 9$$
$$= 34$$

42. C If f(x) is to be a real number, the expression under the radical sign must be zero or greater since the square root of a negative number is not a real number. Therefore, x must be less than or equal to 5 and greater than or equal to -5.

156

43.   C   The product of two even numbers is even. This set is closed under multiplication. The product of two positive numbers is positive, so this set is also closed. The product of two negative numbers is positive. This set, therefore, is not closed under multiplication.

44.   B   53 and 59 are the only prime numbers between 50 and 60.

45.   D $A \cap B = \{2, 3\}$. The union of this with the set
$\{1, 5, 8\} = \{1, 2, 3, 5, 8\}$

46.   A $2^0 = 1$, $2^1 = 2$, $2^2 = 4$. Therefore, $4_{10} = 100_2$.

47.   C   If $a = b^3$, then substitute twice b or 2b for b = $(2b)^3 = 8b^3$. Since $a = b^3$, this = 8a.

48.   D   If a is odd, 3a is also odd, because the product of two odd numbers is odd. Adding 1 to an odd number is the sum of two odd numbers, which is even.

49.   A

$$x^2 - 3x = 28$$
$$x^2 - 3x - 28 = 0 \qquad \text{thus, } (x - 7)(x + 4) = 0$$
$$x - 7 = 0 \qquad\qquad x + 4 = 0$$
$$x = 7 \qquad\qquad x = -4$$

50.   B   The number of committees would be given by:
$$_{10}C_3 = \frac{10 \cdot 9 \cdot 8}{3 \cdot 2 \cdot 1} = 120.$$

51. B   The common solutions are the points where the graphs intersect. The graphs below intersect at two points.

NOTE: The algebraic solution can be found be setting the equations equal to each other:

$$X^2 = 2x + 3$$
$$X^2 - 2x - 3 = 0$$
$$(x - 3)(x + 1) = 0$$
$$x = 3, \quad x = -1$$

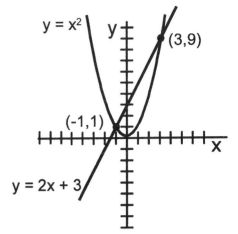

Substituting each x and solving for y: $y = 3^2 = 9$ and $y = (-1)^2 = 1$.

The common points are (3, 9) and (-1, 1).

52. B   A is the set of numbers greater than or equal to 1. B is the set of numbers less than or equal to 1. The only common element is 1.

53. B   By squaring each x and subtracting x from the total you get y. Therefore, the value of y for x = 4 is, $4^2 - 4 = 16 - 4 = 12$.

54. D   If $\sqrt{ab} = 6$, then by squaring both sides, ab = 36. The factors of 36 are: 36 x 1, 18 x 2, 12 x 3, 9 x 4, and 6 x 6. Only 12 can not be found by subtracting a and b.

55. C   If a ≠ 2, it does not necessarily follow that b cannot equal 3. For choice B it is possible that b = 3 under other conditions besides when a = 2. However, it is true that if b ≠ 3 then a cannot equal 2, because if a = 2, then b = 3.

56. D   The shaded portion of the Venn diagram represents the intersection (or common part) of A and B together with the intersection of A and C. In symbols, this is (A ∩ B) ∪ (A ∩ C).

57. D   Since L is under the radical or square root symbol, and g is a constant, to extract a quantity three times larger the quantity under the radical would have to be nine times larger. (The $\sqrt{9} = 3$.)

158

58. D   Since logarithms are really exponents, all three properties are true. The first corresponds to adding exponents in multiplication. The second corresponds to subtracting exponents in division. The third corresponds to raising an exponent to a power.

59. B   Only statement I is false. By substituting numbers you see that $(3 + 4)^2$ does not equal $3^2 + 4^2$. Also, you know that $(a + b)(a + b)$ is equal to $a^2 + 2ab + b^2$, which is not equal to $a^2 + b^2$.

60. D   The square of a number is less than the number itself when the number is a positive fraction less than 1.

61. D   The logic symbols translate, "If not a, then b." For this statement to be false, both the antecedent (a) and the consequent must be false because if b is false, then a is true.

62. A $(4 - 2i)(4 + 2i) = 4^2 - (2i)^2 = 16 - 4i^2$. Since $i = \sqrt{-1}$, then $i^2 = \sqrt{-1} \cdot \sqrt{-1} = -1$. Therefore $16 - 4i^2 = 16 - 4(-1) = 16 + 4 = 20$.

63. D   The slope of a straight line can be found by dividing the difference in the y coordinates by the difference in the x coordinates. In this problem, that would be equal to: $\frac{(a - 5)}{(4 - 3)} = 3$. Solving for a, we get $a = 8$.

64. A   There are 10 favorable numbers divisible by 2: 2, 4, 6, 8, 10, 12, 14, 16, 18, 20. The probability is $\frac{10}{21}$. There are only 4 favorable numbers that are divisible by 5: 5, 10, 15, 20. The probability is $\frac{4}{21}$. The probability of both is $\frac{10}{21} \times \frac{4}{21} = \frac{40}{441}$.

65. A   The log of the product of two numbers is equal to the sum of the logs. (See question 26.) Since $35 = 7 \cdot 5$,
$\log 35 = \log 7 + \log 5 = b + c$.